T0320670

Advanced
Linear Algebra

With an Introduction to
Module Theory

Advanced Linear Algebra

With an Introduction to Module Theory

Shou-Te Chang

National Chung Cheng University, Taiwan

World Scientific

EW JERSEY · LONDON · SINGAPORE · BEIJING · SHANGHAI · HONG KONG · TAIPEI · CHENNAI · TOKYO

Published by

World Scientific Publishing Co. Pte. Ltd.

5 Toh Tuck Link, Singapore 596224

USA office: 27 Warren Street, Suite 401-402, Hackensack, NJ 07601

UK office: 57 Shelton Street, Covent Garden, London WC2H 9HE

Library of Congress Cataloging-in-Publication Data
Names: Chang, Shou-Te, author.
Title: Advanced linear algebra : with an introduction to module theory /
    Shou-Te Chang, National Chung Cheng University, Taiwan.
Description: New Jersey : World Scientific Publishing Co. Pte. Ltd., [2024] | Includes index.
Identifiers: LCCN 2024000487 | ISBN 9789811276354 (hardcover) |
    ISBN 9789811277245 (paperback) | ISBN 9789811276361 (ebook for institutions) |
    ISBN 9789811276378 (ebook for individuals)
Subjects: LCSH: Algebras, Linear--Textbooks.
Classification: LCC QA184.2 .C453 2024 | DDC 512/.5--dc23/eng/20240201
LC record available at https://lccn.loc.gov/2024000487

British Library Cataloguing-in-Publication Data
A catalogue record for this book is available from the British Library.

For any available supplementary material, please visit
https://www.worldscientific.com/worldscibooks/10.1142/13411#t=suppl

Printed in Singapore

Preface

This book is suitable for students who want to go further into any algebraic disciplines. Previously, the author had published two textbooks (both published by World Scientific) with Professor Minking Eie, National Chung Cheng University, Taiwan. In the first book, *A First Course in Linear Algebra*,[1] the reader will familiarize oneself with concepts such as bases, dimension, matrices, and what not. There is a detailed discussion on the spectral theorem and a brief introduction to Cayley-Hamilton theorem and Jordan forms without proof. The topics of the second book, *A Course on Abstract Algebra, 2e*,[2] include introductions to group theory, ring theory and some field theory. For a solid foundation of abstract algebra, the author strongly feels that, in addition to familiarity with contents of the previously mentioned two books (or of other similar textbooks), a third textbook is necessary. The author believes that, equipped with knowledge covered by these three books, a student of mathematics can delve into any algebraic disciplines with complete confidence.

Linear algebra is considered one of the most basic mathematical tools, even for non-mathematicians. However, a student who aspires to a career in mathematics often finds a gaping gap between a first course on linear algebra and the linear algebra needed in graduate schools. For example, an elementary textbook on linear algebra will not provide proofs to results on topics about or related to canonical forms of a square matrix. The proofs would be hard to follow even if they are provided. Another example is

---

[1] ISBN 9789813143111
[2] ISBN 9789813229624

that such a textbook can only focus on finite dimensional vector spaces, while graduate students often have to deal with infinite dimensional vector spaces. These materials are considered too advanced for undergraduates while the same materials are simply assumed in the graduate school. This book fills the need for students who want to fill the gap themselves. It will provide a gentle guidance to motivated students for self-study. It is also a suitable textbook for an advanced undergraduate course on linear algebra or on module theory. Its introduction to sets and modules would be of particular interest to students who aspire to becoming algebraists.

One of the main purposes in this book is the introduction of module theory. The concept of modules is a generalization of the concept of vector spaces. For modules, the scalars only need to form a commutative ring with unity instead of a field. We will find most concepts regarding vector spaces have their counterparts in module theory. Why can't we consider a matrix with entries only in $\mathbb{Z}$? How do we determine whether a matrix over $\mathbb{Z}$ is invertible? And if the inverse matrix exists, how do we find it? We see that many questions answered in linear algebra remain to be explored in module theory. The study of modules opens up a whole new world and new adventures. The results and insights gained in module theory actually also help us further understand linear algebra.

What preparation do we need for using this book? A basic understanding of set theory, group theory and ring theory is preferred. Here is a checklist.

- A mentality for an axiomatic approach.

- Some basic set theory, including the equivalence relation, the well-ordering principle, the pigeonhole principle and the induction principle.

- Definitions of groups, rings, fields and vector spaces. The basic properties of these algebraic structures. Morphisms of each structure and their basic properties.

- Chinese remainder theorem for $\mathbb{Z}$.

- The fundamental theorem of homomorphisms. The three isomorphism theorems.

- The Euclidean domain, the PID (principal ideal domain) and the UFD (factorial domain or unique factorization domain).

A first course on finite dimensional vector spaces is preferable, although not necessary. The treatment of vector spaces in this book will be over any field, instead of just over $\mathbb{R}$ or over $\mathbb{C}$, and so we will start from scratch. However, some mathematical maturity will be very helpful. Besides, inner product spaces and the spectral theorem will not be discussed in this book, and these topics are worth studying.

Basically, there are three parts in this book. The first part is to complete the discussion of vector spaces, bases and dimension. We will discuss the general case of vector spaces of arbitrary dimension over an arbitrary scalar field. To do so, we will need an introduction to cardinality and Zorn's lemma in set theory. We will also introduce the concept of modules during the process. The second part is to complete the discussion of canonical forms for linear endomorphisms and square matrices. For this we will discuss the structure theorem of finitely generated modules over a PID, which also serves as a good entry into module theory. The last part was an introduction to the concept of the tensor product. The tensor product is a useful tool in linear algebra, and an even more useful tool in module theory. These materials will be presented in the categorical spirit without actually spelling out the categorical language in mathematics.

I wish the reader a pleasant and fruitful journey in reading this book. Your input will be greatly appreciated. The author can be reached at `shoute.chang@gmail.com`.

At last, I would like to thank my editor at World Scientific, Ms Kwong Lai Fun for her kindness and assistance. I would also like to thank my students who gave me suggestions and feedback when this book was still just class notes.

Shou-Te Chang
March 31, 2023

About the Author

Shou-Te Chang is currently an Associate Professor at the Department of Mathematics, National Chung Cheng University in Chiayi, Taiwan. She received her Ph.D. in Mathematics in 1993 from University of Michigan, Ann Arbor. Her research interest is in commutative algebra and homological algebra. She has published papers on Horrocks' question, generalized Hilbert-Kunz functions and local cohomology. She also published several undergraduate textbooks on linear algebra and modern algebra with World Scientific.

Contents

Special Notes

Unless otherwise specified, all *rings* are assumed to be commutative with unity (the multiplicative identity). A subring shares the same unity with its super ring. By a *domain* we mean an integral domain, a commutative ring which contains no nontrivial zero divisors.

Let S be any set, we use 1_S to denote the identity map on S.

We will also use the following conventions:

$\mathbb{C}$ = the set of complex numbers.

$\mathbb{R}$ = the set of real numbers.

$\mathbb{Q}$ = the set of rational numbers.

$\mathbb{Z}$ = the set of integers.

$\mathbb{Z}_+ = \{1, 2, 3, 4, 5, \dots\}$, the set of *positive* integers.

$\mathbb{N} = \{0, 1, 2, 3, 4, 5, \dots\}$, the set of *nonnegative* integers, or the set of *natural numbers*.[3]

---

[3] There are two schools of mathematicians in the world, one who views 0 as a natural number and one who doesn't.

CHAPTER 1

Modules and Vector Spaces

In this chapter we will introduce the concept of modules, which is a generalization of the concept of vector spaces. A vector space comes equipped with a field of scalars. For modules, the scalars are only required to be elements from a ring. Certain properties of vectors are still enjoyed by module elements, while some others need modification or are simply lacking for module elements. Surprisingly, to further understand vector spaces and matrices over a field, it is actually better to tackle them as modules or as matrices over a ring.

In this chapter, we will start with defining modules. For modules we may similarly define the concepts of linear combination and linear independence. The main difference between vector spaces and modules is that not every module possesses a basis while a vector space always contains a basis. In a first course on linear algebra, one expects to see why any two bases of a *finite dimensional* vector space contain the same number of elements. In this chapter, we will review an argument for this fact. We will also give the reason why this fact holds for *infinite dimensional* vector spaces. For this, we will discuss some set theory, especially the part regarding *cardinality* of sets.

1.1   Definitions and examples of modules and vector spaces

Linear algebra is the study of vector spaces and matrices. In this chapter we will introduce the more general concept of modules. When we delve into some of the more advanced topics in linear algebra, it is easier to study them from the module angle.

A few words on rings and fields

We say $(R, +, \cdot, 0, 1)$ is a **ring with unity** if it is a set with two binary operations, the addition "+" and the multiplication "$\cdot$", such that it satisfies the following conditions.

(i) The structure $(R, +, 0)$ is an **abelian group**:

  - The binary operation "+" (addition) is associative;
  - The element 0 is the additive identity;
  - Every element in R has an additive inverse;
  - The binary operation "+" is commutative.

(ii) The structure $(R, \cdot, 1)$ is a **monoid**:

  - The binary operation "$\cdot$" (multiplication) is associative;
  - The element 1 is the multiplicative identity.

(iii) The multiplication "$\cdot$" is distributive with respect to the addition "+":

  - $(a + b) \cdot c = a \cdot c + b \cdot c$ for all a, b, $c \in R$;
  - $a \cdot (b + c) = a \cdot b + a \cdot c$ for all a, b, $c \in R$.

The product "$a \cdot b$" is usually abbreviated as "ab". The multiplicative identity "1" is also called the **unity** of R. In some books a ring is not assumed to contain the unity. In this book our rings always contain the unity. If in addition the binary operation $\cdot$ is commutative, we say that R is a **commutative ring**.

In this book, the only rings of interest are the commutative rings with unity. Unless otherwise noted, when we say that R is a **ring** we mean that R is a commutative ring with unity.

If F is a commutative ring with unity which further satisfies the assumption that every nonzero element has a multiplicative inverse, we say that F is a **field**.

In short, a ring is where you can add, subtract and multiply. A field is where you can add, subtract, multiply and divide (by a nonzero element).

Common examples of rings include $\mathbb{Z}$, $\mathbb{Z}_n$ and $R[x]$, the polynomial ring of one variable over the ring R. Common examples of fields include $\mathbb{Q}$, $\mathbb{R}$, $\mathbb{C}$ and Z_p where p is a positive prime integer.

Modules and vector spaces

Let's introduce the modules.

Definition 1.1.1. Let R be a ring. We say that M, or more specifically $(M, +, 0)$, is an **R-module** or a **module over R**[1] if M is an additive group and there is a mapping

$$
\begin{aligned}
R \times M &\longrightarrow M \\
(a, m) &\longmapsto a \cdot m
\end{aligned}
$$

such that

(i) $1 \cdot m = m$,

(ii) $a \cdot (m + n) = a \cdot m + a \cdot n$,

(iii) $(a + b) \cdot m = a \cdot m + b \cdot m$, and

(iv) $a \cdot (b \cdot m) = (ab) \cdot m$

for all $a, b \in R$ and $m, n \in M$.

In the case where $R = F$ is a field, M is also called an **F-vector space** or a **vector space over F**. The elements of F are called **scalars** and the

---

[1] The theory of modules can be established even when R is a non-commutative ring. In that case, we have to differentiate between left and right modules. Presented here is in fact the definition of the left modules.

elements of M are called **vectors**. The addition in M is called the **vector addition**. The element $a \cdot m$ is called the **scalar multiplication** of a and m.

However, for lack of a better term, we will also call $a \cdot m$ the *scalar multiplication* of a and m even when R is not a field. When there leaves no confusion, the element $a \cdot m$ is often abbreviated as am. In fact, we will do just that for the rest of the book.

Always remember that a vector space is also a module. All properties of modules apply to vector spaces.

Next are some examples of modules and vector spaces.

Example 1.1.2. The trivial group $\{0\}$ has a natural module structure. We call it the **trivial** module.

Example 1.1.3. Let R be a ring and n be a positive integer. Let R^n be the direct product ring of n copies of R. Then $(R^n, +)$ is an additive group with a natural scalar multiplication of R on R^n by letting

$$a(r_1, r_2, \ldots, r_n) = (ar_1, ar_2, \ldots, ar_n)$$

for $a, r_1, r_2, \ldots, r_n \in R$. Observe that

$$1(r_1, r_2, \ldots, r_n) = (1r_1, 1r_2, \ldots, 1r_n) = (r_1, r_2, \ldots, r_n);$$

$$a[(r_1, r_2, \ldots, r_n) + (s_1, s_2, \ldots, s_n)] = a(r_1 + s_1, \ldots, r_n + s_n)$$
$$= \big(a(r_1 + s_1), \ldots, a(r_n + s_n)\big) = (ar_1 + as_1, \ldots, ar_n + as_n)$$
$$= (ar_1, ar_2, \ldots, ar_n) + (as_1, as_2, \ldots, as_n)$$
$$= a(r_1, r_2, \ldots, r_n) + a(s_1, s_2, \ldots, s_n);$$

$$(a + b)(r_1, r_2, \ldots, r_n) = \big((a + b)r_1, (a + b)r_2, \ldots, (a + b)r_n\big)$$
$$= (ar_1 + br_1, \ldots, ar_n + br_n) = (ar_1, ar_2, \ldots, ar_n) + (br_1, br_2, \ldots, br_n)$$
$$= a(r_1, r_2, \ldots, r_n) + b(r_1, r_2, \ldots, r_n);$$

$$a\big(b(r_1, r_2, \ldots, r_n)\big) = a(br_1, br_2, \ldots, br_n) = (abr_1, abr_2, \ldots, abr_n)$$
$$= (ab)(r_1, r_2, \ldots, r_n).$$

Hence R^n is an R-module. If F is a field, F^n is an F-vector space. Note that $R = R^1$ may be viewed as an R-module and F an F-vector space.

Example 1.1.4. Let G be an abelian group. Without loss of generality, we might as well assume that G is an additive group. There is a natural way to view G as a $\mathbb{Z}$-module by letting

$$0_{\mathbb{Z}} \cdot a = 0_G,$$
$$k \cdot a = \underbrace{a + \cdots + a}_{k \text{ copies}}, \qquad \text{and}$$
$$(-k) \cdot a = -(k \cdot a) = \underbrace{(-a) + \cdots + (-a)}_{k \text{ copies}}$$

for any $a \in G$ and any positive integer k. The verification that G is thus a $\mathbb{Z}$-module is tedious but straightforward. A question regarding an abelian group may often be translated as a question on a $\mathbb{Z}$-module. Conversely, a $\mathbb{Z}$-module is an additive (abelian) group by definition. Hence, the concepts of $\mathbb{Z}$-modules and abelian groups are basically the same.

Example 1.1.5. Let V be the set of all continuous real-valued functions defined on an interval I of $\mathbb{R}$. There is a natural addition in V defined by

$$(1.1.1) \qquad (f + g)(x) = f(x) + g(x)$$

for f and $g \in V$. For $a \in \mathbb{R}$, there is also a natural scalar multiplication defined by

$$(1.1.2) \qquad (af)(x) = af(x).$$

This follows from the fact (in a course of Advanced Calculus) that $f + g$ and af remain continuous on I. We leave it to the reader to verify that V is an $\mathbb{R}$-vector space.

Similarly, the set of differentiable real-valued functions on an open interval I, or the integrable real-valued functions on a finite closed interval I can both be regarded as $\mathbb{R}$-vector spaces.

Let D be an open region in $\mathbb{R}^m$. Consider the set W of vector fields (vector-valued functions) with D as the domain and $\mathbb{R}^n$ as the codomain. With the vector addition and scalar multiplication defined as in (1.1.1) and (1.1.2), W is an $\mathbb{R}$-vector space.

Example 1.1.6. The space $\mathbb{C}^n$ is a $\mathbb{C}$-vector space. It is also an $\mathbb{R}$-vector space, since the scalar multiplication may be induced by restricting the mapping $\mathbb{C} \times \mathbb{C}^n \to \mathbb{C}^n$ to $\mathbb{R} \times \mathbb{C}^n \to \mathbb{C}^n$.

In general, if F is a subfield of E, then any E-vector space can also be regarded as an F-vector space. If R is a subring of S, then any S-module can be regarded as an R-module.

Example 1.1.7. Let R be a ring and let x_1, x_2, $\ldots$, x_n be indeterminates over R. The ring R may be considered as a subring of $R[x_1, x_2, \ldots, x_n]$. Note that $R[x_1, x_2, \ldots, x_n]$ may also regarded as an R-module.

More generally, if R is a subring of S, then S is an R-module. The scalar multiplication is induced by the product inside S since elements of R are also elements of S. In particular, E is an F-vector space when F is a subfield of E.

Definition 1.1.8. We will use $(a_{ij})_{m \times n}$ to denote an array of m rows and n columns

$$\begin{pmatrix} a_{11} & a_{12} & \cdots & a_{1n} \\ a_{21} & a_{22} & \cdots & a_{2n} \\ \vdots & \vdots & \ddots & \vdots \\ a_{m1} & a_{m2} & \cdots & a_{mn} \end{pmatrix}.$$

The a_{ij}'s are called the **entries** of this array. If all the entries are elements in a ring R, we call it an $m \times n$ **matrix** over R. We will use $M_{m \times n}(R)$ to denote the set of all $m \times n$ matrices over R. An $n \times n$ matrix is called a **square matrix** of size n. We will use $M_n(R)$ to denote the set of square matrices of size n.

We use I_n, or simply I when n is understood, to denote the **identity** matrix of size n. To be precise, $I_n = (a_{ij})_{n \times n}$ where

$$a_{ij} = \text{the Kronecker delta} \delta_{ij} = \begin{cases} 0, & \text{if } i \neq j; \\ 1, & \text{if } i = j. \end{cases}$$

We use $\mathbf{0}_{m \times n}$, or simply $\mathbf{0}$ if m and n are understood, to denote the $m \times n$ matrix whose entries are all 0. This is called the **zero** matrix or the **trivial** matrix of size $m \times n$.

Example 1.1.9. Inside $M_{m \times n}(R)$ there is the usual matrix addition and scalar multiplication:

$$(a_{ij}) + (b_{ij}) = (a_{ij} + b_{ij})$$

$$r(a_{ij}) = (ra_{ij})$$

where a_{ij}, b_{ij} and $r \in R$. These operations make $M_{m \times n}$ into an R-module. The reasoning is similar to that of Example 1.1.3.

Let S and T be sets. We often use T^S to denote the set of all functions from S to T. In general, if S and T are both finite sets, we know that

(1.1.3) $$|T^S| = |T|^{|S|}.$$

Example 1.1.10. Let R be a ring and let S be an arbitrary set. Let f and g be functions in R^S. The function $f + g$ is defined by letting

(1.1.4) $$(f + g)(s) = f(s) + g(s) \qquad \text{for any } s \in S.$$

Let $a \in R$. The function af is defined by letting

(1.1.5) $$(af)(s) = af(s) \qquad \text{for any } s \in S.$$

To verify that R^S is thus an R-module, we need to check the four requirements in Definition 1.1.1. We will demonstrate (iii) and leave the rest of the verifications to the reader. To check that the function $(a + b)f$ equals the function $af + bf$ for any a and $b \in R$ and $f \in R^S$, we compare the values of both functions at every element in the domain S. We have

$$
\begin{aligned}
&\big((a+b)f\big)(s) \\
&= (a+b)f(s), \qquad \text{by (1.1.5)}, \\
&= af(s) + bf(s), \qquad \text{by the distributivity in } R, \\
&= (af)(s) + (bf)(s), \qquad \text{by (1.1.5)}, \\
&= (af + bf)(s), \qquad \text{by (1.1.4)}
\end{aligned}
$$

for all $s \in S$. We conclude that $(a + b)f = af + bf$. Note that the additive identity in R^S the **zero function**, the function sending any element of S to the additive identity 0_R in R.

For example, let $S = \{a, b\}$ be a doubleton set. The following table lists all the elements in $\mathbb{Z}_3^S$:

	f_1	f_2	f_3	f_4	f_5	f_6	f_7	f_8	f_9
$a \mapsto$	0	0	0	1	1	1	2	2	2
$b \mapsto$	0	1	2	0	1	2	0	1	2

This set is a $\mathbb{Z}_3$-vector space of 9 elements. (*Cf.* (1.1.3).) The function f_1 is the trivial vector in $\mathbb{Z}_3^S$. Observe that

$$(f_5 + f_6)(a) = 2 \qquad \text{and} \qquad (f_5 + f_6)(b) = 3 = 0.$$

We have that $f_5 + f_6 = f_7$. Similarly,

$$(2f_8)(a) = 4 = 1 \qquad \text{and} \qquad (2f_8)(b) = 2.$$

We have that $2f_8 = f_6$.

Example 1.1.11. Let R be a ring and let I be an ideal of R. Remember that I is an additive subgroup of R by definition. Moreover, $ra \in I$ for all $r \in R$ and $a \in I$. The product in R induces a scalar multiplication of R on I. Hence I is an R-module. On the other hand, the quotient ring R/I is also an R-module. Remember that R/I remains an additive group, and the scalar multiplication of R on R/I is given by $r \cdot \overline{a} = \overline{ra}$ for $r \in R$ and $\overline{a} \in R/I$.

Remember that R/I is also a ring. Hence R/I is an (R/I)-module as well. For example, $\mathbb{Z}_2 = \mathbb{Z}/2\mathbb{Z}$ is both a $\mathbb{Z}$-module and a $\mathbb{Z}_2$-vector space since $\mathbb{Z}_2$ is a field.

Example 1.1.12. Let R be a ring, I be an ideal of R and M be a module over R/I. Let $r \in R$ and $m \in M$. Define the scalar multiplication $r \cdot m$ to be the scalar multiplication $\overline{r} \cdot m$ over R/I. It is routine to verify that this makes M into an R-module in a most natural way.

Submodules and subspaces

Definition 1.1.13. Let R be a ring and M be an R-module. We say a subset N of M is an R-**submodule** if N is an R-module with the inherited addition and scalar multiplication.

Let F be a field and V an F-vector space. We say a subset W of V is an F-**subspace** if W is an F-vector space with the inherited addition and scalar multiplication.

Remark. We would like to remind the reader that a module or a vector space is by default *nonempty*! Moreover, for a subset N of M to inherit addition and scalar multiplication from M, one should have

- the addition on N is closed, and

- the scalar multiplication of R on N is closed.

We have the following easy result for checking whether N (or W) is a submodule (or a subspace) of M (or V, respectively). Remember that in a vector space, a subspace is a submodule and a submodule is a subspace.

Lemma 1.1.14 (Test for submodules and subspaces). *Let R be a ring and M be an R-module. A subset N of M is a submodule of M if and only if the following three conditions are satisfied:*

(i) $0 \in N$;

(ii) $n + n' \in N$ whenever n and $n' \in N$;

(iii) $an \in N$ whenever $a \in R$ and $n \in N$.

This test applies to subspaces as well.

Proof. The "only if" part: We already have (ii) and (iii) from the Remark to Definition 1.1.13. To show (i), we may find the additive identity n in N since N is a module. We should have $n + n = n$. Hence $0 + n = n = n + n$ in M. We have that $0 = n \in N$ by the cancelation law in M.

The "if" part: Condition (i) guarantees that N is nonempty. Condition (ii) guarantees that there is an inherited addition in N. Condition (iii) further assures that $-n = (-1)n \in N$ for each $n \in N$. (See Exercise 1(c).) Hence N is an additive subgroup of M. Condition (iii) also guarantees that there is an inherited scalar multiplication in N. Conditions (i)–(iv) in Definition 1.1.1 hold in N since they hold true in M. We have shown that N is an R-module. $\square$

Definition 1.1.15. Let M be an R-module. Then $\{0\}$ is clearly a submodule of M, called the **trivial** submodule. The module M itself is also a submodule of M, called the **improper** submodule of M. A submodule N which is properly contained in M is called a **proper** submodule of M.

Example 1.1.16. Note that the three conditions in Lemma 1.1.14 are also used for testing ideals. This is because ideals of a ring R are also submodules of R (see Exercise 3).

Example 1.1.17. Let R be a ring. Then $\{(a,0) \in R^2 : a \in R\}$ is an R-submodule of R^2. The **diagonal** $\{(a,a) \in R^2 : a \in R\}$ is also an R-submodule of R^2.

Example 1.1.18. The set of real-valued continuous functions defined on an open interval I is an $\mathbb{R}$-subspace of $\mathbb{R}^I$ (see Examples 1.1.5 and 1.1.10). The set of real-valued differentiable functions defined on I is an $\mathbb{R}$-subspace of the vector space of real-valued continuous functions defined on I.

Example 1.1.19. Remember that G is an abelian group if and only if it is a $\mathbb{Z}$-module.

Subgroups of an abelian group G are $\mathbb{Z}$-submodules of G since the subgroups are also abelian. Conversely, let H be a $\mathbb{Z}$-submodule of an abelian group G. From Lemma 1.1.14 and Exercise 1(c), we can see that the addition is closed in H and that taking additive inverse is also closed in H since $-h = (-1)h \in H$ for all $h \in H$. Thus H is a subgroup of G. We conclude that a subset of the abelian group G is a subgroup if and only if it is a $\mathbb{Z}$-submodule.

Exercises 1.1

In this set of exercises, F always denotes a field and R a ring.

1. Let M be an R-module. Show that the following assertions are true for all $m \in M$ and for all $a \in R$.

    (a) $a0_M = 0_M$.

    (b) $0_R m = 0_M$. (Normally we will just use the symbol 0 for either 0_R or 0_M when no confusion arises.)

    (c) $(-1)m = -m$. (Here -1 stands for the additive inverse of 1 in R and $-m$ is the additive inverse of m in M.)

    (d) If R is a field and $am = 0$, then $a = 0$ or $m = 0$.

    Give an example to show that the assertion (d) is not necessarily true if R is not a field.

2. Lemma 1.1.14 gives a test for submodules. In this exercise we give an alternative test for submodules.

   Show that a subset N of M is a submodule of M if and only if the following two conditions are satisfied:

   (i) N is nonempty;

   (ii) $an + n' \in N$ whenever $a \in R$ and n, $n' \in N$.

3. Show that the submodules of R are exactly the ideals of R.

4. Show that the set of diagonal square matrices

$$
D = \left\{ \begin{pmatrix} a_1 & 0 & \cdots & 0 \\ 0 & a_2 & \cdots & 0 \\ 0 & 0 & \ddots & 0 \\ 0 & 0 & \cdots & a_n \end{pmatrix}_{n \times n} \in M_n(R) \ : \ a_i \in R \text{ for all } i \right\}
$$

   is an R-submodule of $M_n(R)$.

5. Let $R[x]$ be the polynomial ring of one variable over R and let

$$
\mathscr{P}_n = \{0\} \cup \{f \in R[x] : \deg f \le n\}.
$$

   Show that $\mathscr{P}_n$ is an R-submodule of $R[x]$.

6. Let S be a set and $s_0 \in S$. Show that $\{f \in R^S : f(s_0) = 0\}$ is an R-submodule of R^S.

7. Let M_i, $i = 1, \dots, n$, be R-modules. Remember that the direct product $M_1 \times \cdots \times M_n$ is an additive group. There is also a natural scalar multiplication given by

$$
a(m_1, m_2, \dots, m_n) = (am_1, am_2, \dots, am_n)
$$

   for $a \in R$ and $m_i \in M_i$ for each i. Show that this makes $M_1 \times \cdots \times M_n$ an R-module. This module is often denoted as $M_1 \oplus \cdots \oplus M_n$ and is also called the **direct sum** of the M_i's.

8. Suppose given an index set I and a family of sets $\{X_i\}_{i \in I}$. Define

$$\prod_{i \in I} X_i = \{(x_i)_{i \in I} : x_i \in X_i \text{ for each } i \in I\}$$

and we call it the **direct product** of the X_i's.

Let $\{M_i\}_{i \in I}$ be a family of R-modules. Show that the direct product of R-modules $\prod_{i \in I} M_i$ is an R-module with respect to the natural addition and the scalar multiplication

$$(m_i)_{i \in I} + (m_i')_{i \in I} = (m_i + m_i')_{i \in I};$$
$$a(m_i)_{i \in I} = (am_i)_{i \in I}$$

where $a \in R$ and $(m_i)_{i \in I}, (m_i')_{i \in I} \in \prod_{i \in I} M_i$.

9. Let M_i be an R-module for each $i \in I$. Define $\bigoplus_{i \in I} M_i$ to be the set

$$\{(m_i)_{i \in I} : m_i \in M_i \text{ and } m_i = 0 \text{ except for finitely many } i \text{ in } I\}.$$

We call it the **direct sum** of the M_i's. This is a subset of the direct product $\prod_{i \in I} M_i$. Show that $\bigoplus_{i \in I} M_i$ is an R-submodule of $\prod_{i \in I} M_i$ under the inherited addition and scalar multiplication.

Note that the direct sum and the direct product are the same when I is finite, while the direct product and the direct sum are not the same when I is infinite. For example, $(1, 1, 1, \dots)$ is an element in $\prod_{i=1}^{\infty} \mathbb{Z}$ but not in $\bigoplus_i^{\infty} \mathbb{Z}$.

10. Let M be an R-module and let M_i be a submodule of M for each $i \in I$. Show that $\bigcap_{i \in I} M_i$ is a submodule of M. Give an example to show that $\bigcup_{i \in I} M_i$ is not necessarily a submodule of M.

11. Let M be an R-module and let M_1 and M_2 be two submodules of M. Show that $M_1 \bigcup M_2$ is a submodule of M if and only if $M_1 \subseteq M_2$ or $M_2 \subseteq M_1$.

12. Let M be an R-module and let M_1, M_2 and N be submodules of M. Show that $N \subseteq M_1$ or $N \subseteq M_2$ if $N \subseteq M_1 \cup M_2$.

13. Let $(M, +)$ be an additive (abelian) group. Show that the scalar multiplication given in Example 1.1.4 is the only mapping that makes M into a $\mathbb{Z}$-module.

14. Let V be a $\mathbb{Q}$-vector space. Show that the given scalar multiplication is the only mapping that makes $(V, +)$ into a $\mathbb{Q}$-vector space.

15. Let M be a non-trivial finite abelian group. Is it possible to make M into a $\mathbb{Q}$-vector space?

16. Let M be an R-module and let S be a set. Construct a natural addition in M^S and a natural scalar multiplication by elements of R, so that M^S becomes an R-module.

17. Let $\varphi \colon R \to S$ be a ring homomorphism and let M be an S-module. Show that M is also an R-module if we define the scalar multiplication by letting $a \cdot m = \varphi(a) \cdot m$ for $a \in R$ and $m \in M$. (Example 1.1.12 is a special case of this result.)

18. Let M be an R-module.

   (a) Let I be an ideal of R such that $am = 0$ for all $a \in I$ and $m \in M$. Show that M may be made into an R/I-module in a natural way.

   (b) Define
   $$\operatorname{Ann}_R M = \{a \in R : am = 0 \text{ for all } m \in M\}.$$

   This is called the **annihilator** of M in R. Show that $\operatorname{Ann}_R M$ is an ideal of R. Hence M is also an $(R/\operatorname{Ann}_R M)$-module. When R is understood, we may simply write $\operatorname{Ann} M$ for $\operatorname{Ann}_R M$.

19. For this problem, we will assume that the ring R is not necessarily commutative. In this case Definition 1.1.1 defines what is called a **left R-module** or a **left module** over R.

   We say that an additive group M is a **right R-module** or a **right module** over R if there is a mapping
   $$\begin{aligned} M \times R &\longrightarrow M \\ (m, a) &\longmapsto m \cdot a \end{aligned}$$
   such that

   (i) $m \cdot 1 = m$,

(ii) $(m + n) \cdot a = m \cdot a + n \cdot a$,

(iii) $m \cdot (a + b) = m \cdot a + m \cdot b$, and

(iv) $(m \cdot b) \cdot a = m \cdot (ab)$

for all a, $b \in R$ and m, $n \in M$.

Now let R be a non-commutative ring and let M be a left R-module. Define the right scalar multiplication as the mapping

$$
\begin{array}{ccc}
M \times R & \longrightarrow & M \\
(m,\, a) & \longmapsto & m \cdot_r a = a^{-1} \cdot_l m.
\end{array}
$$

Show that $\cdot_r$ makes M a right R-module.

1.2   Linear combinations and linear independence

In this section, we discuss two important concepts for vector spaces. These two concepts also apply to modules.

Linear combinations

Definition 1.2.1. Let M be an R-module and let m_1, m_2, ..., m_n be elements in M. We say an element of the form

$$
\sum_{i=1}^{n} a_i m_i, \qquad a_i \in R,
$$

is a **linear combination** of m_1, m_2, ..., m_n over R. By convention, we define *the empty sum* to be the zero vector. By the empty sum we mean the sum of adding "nothing" together.

Example 1.2.2. Consider $u = (1, -1, 3)$, $v = (1, -3, 4)$ and $w = (1, 1, 2)$ in $\mathbb{Z}^3$. Is u a linear combination of v and w over $\mathbb{Z}$? How about over $\mathbb{Q}$?

Solution. For this example we need to find a, $b \in \mathbb{Z}$ such that

$$
(1, -1, 3) = a(1, -3, 4) + b(1, 1, 2).
$$

If we solve the equation over $\mathbb{Q}$, we will see that $a = b = 1/2$ is the only solution. Hence $(1, -1, 3)$ is not a linear combination of v and w over $\mathbb{Z}$. However, it *is* a linear combination of v and w over $\mathbb{Q}$.   ◇

Example 1.2.3. Is $\bar{1}$ a $\mathbb{Z}$-linear combination of $\bar{6}$ and $\bar{8}$ in $\mathbb{Z}_{39}$?

Solution. Since $(-1)6 + 8 = 2$, we have $(-20)6 + (20)8 = 40$. Thus

$$(-20)\bar{6} + (20)\bar{8} = \overline{40} = \bar{1}$$

in $\mathbb{Z}_{39}$. Yes! The element $\bar{1}$ is a $\mathbb{Z}$-linear combination of $\bar{6}$ and $\bar{8}$ over $\mathbb{Z}$. ◇

Generating sets and linear spans

Let S be a subset of M. Exercise 10 in §1.1 tells us that the intersection of all submodules containing S inside M is a submodule of M. Clearly it is *the* smallest submodule containing S in M.

Definition 1.2.4. Let M be an R-module and let S be a subset of M. We use $\langle S \rangle$ to denote the smallest submodule containing S inside M. If $S = \{m_1, m_2, \ldots, m_n\}$, we will simply write $\langle m_1, m_2, \ldots, m_n \rangle$ for $\langle S \rangle$.

If $M = \langle S \rangle$, we say that S **generates** or **spans** M over R. We can also say that S is a **generating set** for M over R. If M can be generated by a finite set over R, we say that M is a **finitely generated** R-module.

If M is a vector space over the field F, it is also customary to use $\mathrm{Sp}_F S$ (or $\mathrm{Sp}\, S$ if F is understood) instead of $\langle S \rangle$ to denote the subspace generated by S. We call $\mathrm{Sp}_F S$ the F-**linear span** of S.

Remark. It is easy to see that $\langle \varnothing \rangle = \langle 0 \rangle = \{0\}$ is the **trivial** module.

Lemma 1.2.5. *Let M be an R-module. Let N be a submodule of M. Let S and T be subsets of M. The following statements are true.*

(a) *Let N be a submodule of M. Then $S \subseteq N$ if and only if $\langle S \rangle \subseteq N$.*

(b) *If $S \subseteq T$, then $\langle S \rangle \subseteq \langle T \rangle$.*

(c) *Let $m \in M$. Then $m \in \langle S \rangle$ if and only if $\langle S \rangle = \langle S \cup \{m\} \rangle$.*

Proof. (a) This follows from Definition 1.2.4 that $\langle S \rangle$ is the smallest submodule of M which contains S.

(b) Since $S \subseteq T \subseteq \langle T \rangle$, the result follows form (a).

(c) The "if" part: This is true since $m \in S \cup \{m\} \subseteq \langle S \cup \{m\} \rangle = \langle S \rangle$.

The "only if" part: It remains to verify $\langle S \rangle \supseteq \langle S \cup \{m\} \rangle$ from (b). Since $m \in \langle S \rangle$, we have that $\langle S \rangle \supseteq S \cup \{m\}$. The result follows from (a). $\square$

The characterization in Definition 1.2.4 does not describe which elements are in $\langle S \rangle$. We need the following result to better understand $\langle S \rangle$.

Proposition 1.2.6. *Let S be a subset of an R-module M. Then $\langle S \rangle$ consists of all R-linear combinations of elements in S. Specifically, an element is in $\langle S \rangle$ if and only if it is of the form $a_1 m_1 + \cdots + a_n m_n$ where $a_i \in R$ and $m_i \in S$.*

Proof. Let N be the set of R-linear combinations of elements of S. We will show that $N = \langle S \rangle$.

"$\subseteq$": Clearly any linear combination of elements in S are in $\langle S \rangle$ from Lemma 1.1.14.

"$\supseteq$": Let $m \in S$. The element $m = 1m$ is an R-linear combination of elements of S. Hence $m \in N$. This shows that $S \subseteq N$. The empty sum is defined to be 0, and so $0 \in N$ by default. It is also easy to see that the sum of two R-linear combinations of elements in S is an R-linear combination of elements in S. So is a scalar multiplication of an R-linear combination of elements in S. Lemma 1.1.14 tells us that N is a submodule containing S in M. It follows from Lemma 1.2.5(a) that $\langle S \rangle \subseteq N$. $\square$

Thanks to Proposition 1.2.6, it is customary to denote $\langle m_1, m_2, \ldots, m_n \rangle$ by $Rm_1 + Rm_2 + \cdots + Rm_n$ or even $m_1 R + m_2 R + \cdots + m_n R$.[2] In particular, when $M = \langle m \rangle = Rm$, we say M is a **cyclic** module.

Example 1.2.7. Let R be a ring. As an R-module, R itself is cyclic since $R = R1$. Let I be an ideal of R. The ideal I is a cyclic R-module if and only if it is a principal ideal. The module R/I is a cyclic module generated by $\overline{1}$ over R as well as over R/I.

Example 1.2.8. (a) Is $\mathbb{Z}_{57} \times \mathbb{Z}_8$ a cyclic $\mathbb{Z}$-module?

---

[2]This is rather the notation for right modules.

(b) Is $\mathbb{Z}_{57} \times \mathbb{Z}_{81}$ a cyclic $\mathbb{Z}$-module?

Solution. At this point we have no adequate tools for this problem yet, and so we will use brutal force for now. However, since we have mentioned in Example 1.1.4 that $\mathbb{Z}$-modules may be viewed as abelian groups, we may use *Chinese remainder theorem* (CRT) for $\mathbb{Z}$ to give us a hint.

(a) From CRT we have that $\mathbb{Z}_{57} \times \mathbb{Z}_8 \simeq \mathbb{Z}_{57 \cdot 8}$ is cyclic as a group. By reviewing the proof of CRT, we can see that $(1,1)$ is a generator. To view this as a $\mathbb{Z}$-module, observe that $57 - 7 \cdot 8 = 1$. We have

$$(\overline{1}, \overline{0}) = -56(\overline{1}, \overline{1})$$
$$(\overline{0}, \overline{1}) = 57(\overline{1}, \overline{1})$$

in $\mathbb{Z}_{57} \times \mathbb{Z}_8$, and

$$(\overline{m}, \overline{n}) = m(\overline{1}, \overline{0}) + n(\overline{0}, \overline{1}) = (-56m + 57n)(\overline{1}, \overline{1})$$

for $m,\, n \in \mathbb{Z}$. Thus, $\mathbb{Z}_{57} \times \mathbb{Z}_8$ is a cyclic module generated by $(\overline{1}, \overline{1})$ over $\mathbb{Z}$.

(b) From CRT, we conjecture that $\mathbb{Z}_{57} \times \mathbb{Z}_{81}$ is not a cyclic group, since 57 and 81 are not relatively prime.

Suppose $M = \mathbb{Z}_{57} \times \mathbb{Z}_{81}$ is a cyclic $\mathbb{Z}$-module. Find a generator $(\overline{k}, \overline{\ell})$ for M over $\mathbb{Z}$. Let $m = 3^4 \cdot 19$, which is the l.c.m. of 57 and 81. Clearly, $m < |M| = 57 \cdot 81 = 3^5 \cdot 19$. Observe that $m(\overline{k}, \overline{\ell}) = (\overline{0}, \overline{0})$. This implies that $(\overline{k}, \overline{\ell})$ cannot generate more than m elements in M, a contradiction. To conclude, M is not cyclic over $\mathbb{Z}$.   ◇

Example 1.2.9. Let R be a ring and $R[x]$ be the polynomial ring of one variable over R. The set $S = \{1, x, x^2, x^3, \dots\}$ generates $R[x]$ over R, since all polynomials are R-linear combinations of the monomials in S. It is also clear that $\mathscr{P}_n$ (see Exercise 5, §1.1) is generated by $\{1, x, x^2, \dots, x^n\}$ over R.

Example 1.2.10. Let R be a ring. We define the element e_i in R^n to be the row whose i-th entry is 1 and whose other entries are all 0:

$$e_i = \quad (0, \dots, 0, 1, 0, \dots, 0)$$
$$\uparrow$$
$$\text{the } i\text{-th position}$$

Then R^n is generated by $\{\, e_1, e_2, \ldots, e_n \,\}$ over R. In fact,

$$(a_1, a_2, \ldots, a_n) = a_1 e_1 + \cdots + a_n e_n.$$

Example 1.2.11. Let R be a ring and m, n be positive integers. Define e_{ij} to be the matrix in $M_{m \times n}(R)$ whose (i,j)-entry is 1 and whose other entries are all 0. Then $M_{m \times n}(R)$ is generated by

$$\{\, e_{ij} \in M_{m \times n}(R) : i = 1, \ldots, m, \ j = 1, \ldots, n \,\}$$

over R. Indeed, an arbitrary matrix

$$\left(a_{ij} \right)_{m \times n} = \sum_{i,j} a_{ij} e_{ij}$$

is a linear combination of the e_{ij}'s over R.

Linear independence

Definition 1.2.12. Let R be a ring and let M be an R-module. We say that m_1, m_2, $\ldots$, m_n are **linearly independent** over R if for $a_1, \ldots, a_n$ in R we have

$$a_1 m_1 + \cdots + a_n m_n = 0 \quad \Longrightarrow \quad a_1 = \cdots = a_n = 0.$$

Otherwise, we say that m_1, m_2, $\ldots$, m_n are **linearly dependent over** R.

The empty set is by default linearly independent over R. A finite subset $\{m_1, m_2, \ldots, m_n\}$ of M is said to be linearly independent over R if m_1, m_2, $\ldots$, m_n are linearly independent over R. An infinite set is said to be linearly independent over R if all of its finite subsets are linearly independent over R. An arbitrary set is linearly dependent over R if it is not linearly independent over R.

A relation of the form

$$0 m_1 + 0 m_2 + \cdots + 0 m_n = 0$$

is, not surprisingly, called a *trivial* relation among m_1, m_2, $\ldots$, m_n. On the contrary, if we can find $a_1, a_2, \ldots, a_n$ in R which are not all zero such that

$$a_1 m_1 + a_2 m_2 + \cdots + a_n m_n = 0,$$

it is called a *non-trivial* relation among $m_1, m_2, \ldots, m_n$. Definition 1.2.12 basically says that a set is linearly dependent if and only if we can find *one* non-trivial relation among finitely many elements in the given set.

We have the following immediate results.

Lemma 1.2.13. *Let M be a module over the ring R. Let S and T be subsets of M such that $S \subseteq T$. The following statements are true.*

(a) *If T is linearly independent over R, then so is S.*

(b) *If S is linearly dependent over R, then so is T.*

Proof. (b) This follows from the fact that any non-trivial relation (among finitely many elements) in S is also a non-trivial relation in T.

The statement (a) is equivalent to (b). $\qquad\qquad\qquad\qquad\qquad\qquad\square$

Example 1.2.14. In $\mathbb{Z}^2$, the elements $(2,3)$ and $(3,-5)$ are linearly independent over $\mathbb{Z}$. This is true since

$$m(2,3) + n(3,-5) = (0,0)$$

does not have any non-trivial solution in $\mathbb{Q}$, let alone in $\mathbb{Z}$.

Example 1.2.15. Let R be a ring. In R^n, $e_1, e_2, \ldots, e_n$ form a linearly independent set over R. Let $a_1, a_2, \ldots, a_n \in R$ be such that

$$(0,\ldots,0) = \sum_{i=1}^{n} a_i e_i = (a_1, a_2, \ldots, a_n).$$

Then $a_1 = a_2 = \cdots = a_n = 0$.

Example 1.2.16. The set

$$\{e_{ij} \in M_{m\times n}(R) : i = 1, 2, \ldots, m, \ j = 1, 2, \ldots, n\}$$

is R-linearly independent in $M_{m\times n}(R)$. Let $a_1, a_2, \ldots, a_n \in R$ be such that

$$0 = \mathbf{0}_{m\times n} = \sum_{\substack{i=1,\ldots,m \\ j=1,\ldots,n}} a_{ij} e_{ij} = \begin{pmatrix} a_{11} & a_{12} & \cdots & a_{1n} \\ a_{21} & a_{22} & \cdots & a_{2n} \\ \vdots & \vdots & \ddots & \vdots \\ a_{m1} & a_{m2} & \cdots & a_{mn} \end{pmatrix}.$$

Then $a_{ij} = 0$ for all i and j.

Example 1.2.17. Let n be an integer ≥ 2. Consider $\mathbb{Z}_n$ as a $\mathbb{Z}_n$-module. We claim that $\overline{k}$ is linearly independent over $\mathbb{Z}_n$ if and only if $(n, \ k) \sim 1$.

Suppose $(n, k) \sim 1$. Let $m \in \mathbb{Z}$ be such that $\overline{m}\,\overline{k} = \overline{0}$. This implies that $n \mid mk$ in $\mathbb{Z}$. Since n and k are relatively prime with each other, we have $n \mid m$ and $\overline{m} = \overline{0}$. We conclude that $\overline{k}$ is linearly independent over $\mathbb{Z}_n$.

Conversely, suppose $d \sim (n, k) \not\sim 1$. Then $\overline{(n/d)}\,\overline{k} = \overline{0}$. Since $\overline{n/d} \neq \overline{0}$ in $\mathbb{Z}_n$, this shows that $\overline{k}$ is linearly dependent over $\mathbb{Z}_n$.

On the other hand, consider $\mathbb{Z}_n$ as a $\mathbb{Z}$-module. No element in $\mathbb{Z}_n$ is linearly independent over $\mathbb{Z}$. Note that $n\overline{k} = \overline{0}$ is a nontrivial relation for any $k \in \mathbb{Z}$.

From the examples above we can see that the concept of linear independence is more complicated for modules. For vector spaces, the situation is more straightforward.

Proposition 1.2.18. *Let F be a field and let V be an F-vector space.*

(a) *Any nonzero vector is linearly independent over F.*

(b) *The set $\{v_1, \ v_2, \ \ldots, \ v_n\}$ is linearly independent over F if and only if*

  - $v_1 \neq 0$, *and*
  - $v_i \notin \mathrm{Sp}\{v_1, \ v_2, \ \ldots, \ v_{i-1}\}$ *for $i = 2, \ldots, n$.*

(c) *The set $\{v_1, \ v_2, \ \ldots, \ v_n\}$ is linearly independent over F if and only if*

  - $\{v_1, \ v_2, \ \ldots, \ v_{n-1}\}$ *is linearly independent over F, and*
  - $v_n \notin \mathrm{Sp}\{v_1, \ v_2, \ \ldots, \ v_{n-1}\}$.

Proof. (a) See Exercise 1(d), §1.1.

(b) Let $S = \{v_1, \ v_2, \ \ldots, \ v_n\}$.

The "only if" part: If $v_1 = 0$, then $1v_1 = 0$, and this says that S is linearly dependent over F. If $v_i \in \mathrm{Sp}\{v_1, \ v_2, \ \ldots, \ v_{i-1}\}$ for some $i \geq 2$, then

$$v_i = a_1 v_1 + \cdots + a_{i-1}v_{i-1}, \qquad a_1, \ldots, a_{i-1} \in R.$$

This gives a non-trivial relation

$$a_1 v_1 + \cdots + a_{i-1}v_{i-1} - 1v_i = 0$$

in S. Hence S is linearly dependent over F.

The "if" part: Suppose S is linearly dependent over F. There is a non-trivial relation among the elements in S. If $v_1 = 0$, we are done. If $v_1 \neq 0$, this relation must involve vectors other than v_1 by part (a). Let this nontrivial relation be

$$a_1 v_1 + a_2 v_2 + \cdots + a_i v_i = 0$$

where $a_i \neq 0$ for some i such that $2 \leq i \leq n$. This implies that

$$v_i = -a_i^{-1}(a_1 v_1 + a_2 v_2 + \cdots + a_{i-1} v_{i-1}) \in \mathrm{Sp}\{v_1, v_2, \ldots, v_{i-1}\}.$$

(c) From (b), the set $\{v_1, v_2, \ldots, v_{n-1}\}$ is linearly independent over F and $v_n \notin \mathrm{Sp}\{v_1, v_2, \ldots, v_{n-1}\}$ if and only if

- $v_1 \neq 0$,

- $v_i \notin \mathrm{Sp}\{v_1, v_2, \ldots, v_{i-1}\}$ for $i = 2, \ldots, n-1$, and

- $v_n \notin \mathrm{Sp}\{v_1, v_2, \ldots, v_{n-1}\}$.

From (b) again, these conditions hold if and only if $\{v_1, v_2, \ldots, v_n\}$ is linearly independent over F. □

Hence, to build a linearly independent set in a vector space, we start with a nontrivial vector, then we pick consecutive elements which are not in the span of the previously chosen elements. Note that this is not true for R-modules in general as we can see in the next example.

Example 1.2.19. The set $\{2, 3\}$ is clearly linearly dependent over $\mathbb{Z}$ since

$$3 \cdot 2 + (-2) \cdot 3 = 0.$$

However, $3 \notin \mathbb{Z}2$, which is the module of even integers. This example shows that Proposition 1.2.18 does not apply to modules.

Example 1.2.20. Let x be an indeterminate over R. If we construct an R-linear combination of finitely many monomials, we have a polynomial. The polynomial is 0 if and only if all the coefficients are 0. Hence, the set $\{1, x, x^2, x^3, \ldots\}$ is linearly independent over R in $R[x]$. In particular, $\{1, x, x^2, \ldots, x^n\}$ is linearly independent over R for all n.

Exercises 1.2

Throughout these exercises R denotes a ring.

1. Consider the $\mathbb{R}$-vector space $\mathbb{R}^{\mathbb{R}}$. Are the two functions $\cos x$ and $\sin x$ linearly independent over $\mathbb{R}$?

2. Let S be a set and let $s \in S$. Define $\chi_s \in R^S$ to be the function

$$\chi_s(t) = \begin{cases} 1, & \text{if } t = s; \\ 0, & \text{otherwise.} \end{cases}$$

This is called the **characteristic function at** s.

Show that $\{\chi_s\}_{s \in S}$ is a linearly independent subset in R^S. Show that $R^S = \mathrm{Sp}\{\chi_s\}_{s \in S}$ if S is a finite set. Is this assertion still true if S is infinite?

3. Let I be an index set and let M and N be submodules of the R-module L. Define

$$M + N = \{m + n \in L : m \in M, \ n \in N\}$$

and we call it the **sum** of M and N.

   (a) Show that $M + N$ is a submodule of L.

   (b) Show that $M + N$ is the smallest submodule in L containing both M and N, that is, $M + N = \langle M \cup N \rangle$.

4. Let M_i be a submodule of the R-module L for each $i \in I$. Define

$$\sum_{i \in I} M_i = \{m_{i_1} + \cdots + m_{i_n} \in L : n \in \mathbb{N}, \ m_{i_j} \in M_{i_j} \text{ for all } j\}.$$

We call it the **sum** of the M_i's. Show that $\sum_{i \in I} M_i$ is a submodule of L.

With this terminology, Proposition 1.2.6 tells us that $\langle S \rangle = \sum_{m \in S} Rm$.

5. Let L, M and N be submodules of an R-module.

(a) Is the relation

$$L \cap (M + N) = (L \cap M) + (L \cap N)$$

true? If not, give a counterexample.

(b) Show that

$$L \cap (M + (L \cap N)) = (L \cap M) + (L \cap N).$$

6. Let $M_1, M_2, \ldots, M_n$ be submodules of the R-module M such that $M = M_1 + \cdots + M_n$. We say that $M_1, M_2, \ldots, M_n$ are **independent** over R if whenever $m_1 + \cdots + m_n = 0$ where $m_i \in M_i$ for each i, we have that $m_1 = \cdots = m_n = 0$. We will use

$$M = M_1 \oplus \cdots \oplus M_n$$

to denote the fact that $M = M_1 + \cdots + M_n$ and $M_1, \ldots, M_n$ are independent over R. We will call M the **(internal) direct sum** of $M_1, M_2, \ldots, M_n$.

Let $m_1, m_2, \ldots, m_n$ be elements in an R-module.

(a) Suppose $m_1, m_2, \ldots, m_n$ are linearly independent over R. Show that $Rm_1, Rm_2, \ldots, Rm_n$ are independent over R.

(b) Suppose that $Rm_1, Rm_2, \ldots, Rm_n$ are independent over R. Is it true that $m_1, m_2, \ldots, m_n$ are linearly independent over R?

7. Can you find a countable generating set for $\bigoplus_{i=1}^{\infty} R$?

8. Let V be a vector space over the field F and let S be a (not necessarily finite) subset of V. Let $v \in V$. Show that $S \cup \{v\}$ is linearly independent over F if and only if S is linearly independent over F and $v \notin \operatorname{Sp} S$. This is a generalization of Proposition 1.2.18(c).

1.3   Bases

The concept of basis plays an essential role in the study of vector spaces. In this section we will examine whether this concept also makes sense for modules.

Bases

Definition 1.3.1. In an R-module, we say a set B is a **base**, **basis** or **free basis** for M over R if

- B generates M, and

- B is linearly independent over R.

Proposition 1.3.2. *Every element in a module can be expressed uniquely as a linear combination of the base elements.*

Proof. Let B be a basis for M over R and let $m \in M$. Since B generates M, m is a R-linear combination of finitely many elements in B. In other words, we can write $m = \sum_{f \in B} a_f f$ such that $a_f \in R$ for all $f \in B$ and $a_f = 0$ except for finitely many f. Suppose we can find another expression $m = \sum_{f \in B} b_f f$ where $b_f \in R$ for all $f \in B$ and $b_f = 0$ except for finitely many f's. Then

$$\sum_{f \in B} (a_f - b_f) f = 0.$$

By the linear independence of the elements in B, $a_f - b_f = 0$ for all $f \in B$. This implies that $a_f = b_f$ for $f \in B$. $\square$

Definition 1.3.3. If an R-module M has an R-basis, we say that M is a **free** R-module or a free module over R.

Not every module is free. Not every module has a basis.

Example 1.3.4. In $\mathbb{Z}_5$, no element is linearly independent over $\mathbb{Z}$ since $5\overline{k} = \overline{0}$ for all $\overline{k} \in \mathbb{Z}_5$. Hence, no nonempty subset of $\mathbb{Z}_5$ is linearly independent over $\mathbb{Z}$. We conclude that there are no bases for $\mathbb{Z}_5$ over $\mathbb{Z}$.

Example 1.3.5. Let R be a ring.

(1) The set $\{e_1, e_2, \ldots, e_n\}$ is an R-basis for R^n. See Examples 1.2.10 and 1.2.15. This is called the *standard basis* of R^n over R.

(2) The set $\{e_{ij} : i = 1, 2, \ldots, m, \ j = 1, 2, \ldots, n\}$ is an R-basis for $M_{m \times n}(R)$. See Examples 1.2.11 and 1.2.16. This is the standard basis of $M_{m \times n}(R)$ over R.

(3) Let x be an indeterminate over R. The set $\{1, x, x^2, x^3, x^4, \dots\}$ is an R-basis for $R[x]$. See Examples 1.2.9 and 1.2.20. This is the standard basis of $R[x]$ over R. Similarly, the set $\{1, x, x^2, \dots, x^n\}$ is called the standard basis of $\mathscr{P}_n$ over R.

(4) Let S be a finite set. The set $\{\chi_s : s \in S\}$ is an R-basis for R^S. This is the standard basis of R^S over R. Note that

$$f = \sum_{s \in S} f(s)\chi_s, \qquad \text{for } f \in R^S.$$

See Exercises 2, §1.2.

Posets

Before we give a characterization to the concept of bases, we need to introduce the concept of "order" or "partial order".

Definition 1.3.6. Let $(S, \le)$ be a set with a relation $\le$. We say $(S, \le)$ is a **partially ordered set** or a **poset** if the following conditions are satisfied for all a, b, $c \in S$:

(i) (Reflexivity) $a \le a$;

(ii) (Transitivity) if $a \le b$ and $b \le c$ then $a \le c$;

(iii) (Antisymmetry) if $a \le b$ and $b \le a$ then $a = b$.

The relation $\le$ is called a **partial order**.

We use the descriptive adjective "partial" because two elements need not be comparable.

Example 1.3.7. Let S be an arbitrary set and let $\mathscr{P}(S)$ be the power set of S (the set of all the subsets of S). Then $(\mathscr{P}(S), \subseteq)$ is a poset.

Example 1.3.8. Let S be a set. We may also define a relation $\preceq$ on $\mathscr{P}(S)$ by letting $A \preceq B$ if and only if $B \subseteq A$. Then $(\mathscr{P}(S), \preceq)$ is a poset. A partial order is not only about size.

Example 1.3.9. Let $\leq$ be the relation "less than or equal to" in the usual sense. Let $|$ denote the relation of divisibility, that is, $a|b$ if and only if $b = ac$ for some c. Then $(\mathbb{Z}_+, \leq)$ and $(\mathbb{Z}_+, |)$ are both posets. However, note that $(\mathbb{Z}, \leq)$ is a poset while $(\mathbb{Z}, |)$ is not! The relation $|$ does not satisfy antisymmetry in $\mathbb{Z}$. For example, we have $1|-1$ and $-1|1$ but $1 \neq -1$ in $\mathbb{Z}$.

Definition 1.3.10. Let $(S, \leq)$ be a poset and let T be a subset of S.

Let $u \in S$. We say u is an **upper bound** for T if $a \leq u$ for all $a \in T$. Similarly, we say l is a **lower bound** for T if $l \leq a$ for all $a \in T$.

Let M and m be in T. We say M is a **maximal** element of T if no element of T is greater than M. In other words, if $a \in T$ and $M \leq a$ then $M = a$. We say m is a **minimal** element of T if no element of T is less than m. In other words, if $a \in T$ and $a \leq m$ then $m = a$.

Let g and l be in T. We say that g is *the* **greatest** element in T if $a \leq g$ for all $a \in T$. We say that l is *the* **least** element in T if $l \leq a$ for all $a \in T$.

Remark. One may notice the word "the" in the phrases "the greatest" and "the least". The word "the" in mathematics means "the unique" or "one and only". Suppose g and h are both greatest elements in a subset T within a poset S. Then $g \leq h$ and $h \leq g$ by definition. This implies that $g = h$ by antisymmetry.

Example 1.3.11. Consider the poset $(\mathbb{Z}, \leq)$ in Example 1.3.9. There are neither maximal nor minimal elements in it.

In this example, we can see that maximal or minimal elements may not exist in a poset.

Example 1.3.12. Let $S = \{1, 2, 3, 4\}$. In the poset $(\mathscr{P}(S), \subseteq)$, the empty set $\varnothing$ is the only minimal element; it is also the least element in $\mathscr{P}(S)$. Similarly, the set S is the greatest element and the only minimal element in $\mathscr{P}(S)$.

Consider the poset $(\mathscr{A}, \subseteq)$ where

$$\mathscr{A} = \{\{1\}, \{3\}, \{1,2\}, \{2,3\}, \{2,4\}, \{2,3,4\}\} \subseteq \mathscr{P}(S).$$

The elements $\{1\}$, $\{3\}$ and $\{2,4\}$ are minimal and the elements $\{1,2\}$ and $\{2,3,4\}$ are maximal. There are neither least nor greatest elements in $\mathscr{A}$.

In this example, we can see that in a poset: (i) maximal elements (or minimal elements respectively) may not be unique even when they exist, and (ii) a minimal (or maximal) element may not be the least (or greatest, respectively) element.

We are now ready to give an important characterization for bases of a vector space.

Proposition 1.3.13. *Let F be a field and V be an F-vector space. The following conditions for a subset B of V are equivalent.*

(i) *The subset B is a basis for V over F.*

(ii) *The subset B is a maximal F-linearly independent subset in V.*

(iii) *The subset B is a minimal generating set for V over F.*

Proof. "(i) $\Rightarrow$ (ii)": Let B be a basis and let C be a linearly independent subset containing B. Suppose C properly contains B. Find $v \in V$ such that $v \in C \setminus B$. Since B is a basis, we have $v \in \operatorname{Sp} B$. Find $v_1, v_2, \ldots, v_n \in B$ such that $v = a_1 v_1 + \cdots + a_n v_n$ where $a_1, a_2, \ldots, a_n \in F$. This gives a non-trivial relation $1v - a_1 v_1 - \cdots - a_n v_n = 0$ in C, a contradiction. Hence B is a maximal linearly independent subset in V.

"(ii) $\Rightarrow$ (iii)": Let B be a maximal linearly independent subset in V. First we show that B is a generating set. Suppose not. Find $v \in V$ such that $v \notin \operatorname{Sp} B$. Then $B \cup \{v\}$ is linearly dependent according to the maximality of B. We may find $v_1, v_2, \ldots, v_n \in B$ and $a, a_1, a_2, \ldots, a_n \in F$ such that $av + a_1 v_1 + \cdots + a_n v_n = 0$ is a nontrivial relation. Note that $a \neq 0$. Otherwise we would have a nontrivial relation in B, a contradiction to B being linearly independent. Thus we have

$$v = a^{-1}(-a_1 v_1 - \cdots - a_n v_n) \in \operatorname{Sp} B,$$

a contradiction again. We conclude that $V = \operatorname{Sp} B$.

If B is not a minimal generating set, B contains a proper subset C such that C is a generating set for V. Find $v \in B \setminus C$. Since $v \in V = \operatorname{Sp} C$, we may find $v_1, v_2, \ldots, v_n \in C$ such that

$$v = a_1 v_1 + \cdots + a_n v_n, \qquad \text{for } a_1, a_2, \ldots, a_n \in F.$$

This gives a nontrivial relation $1v - a_1 v_1 - \cdots - a_n v_n = 0$ in B, a contradiction to the linear independency of B. We have shown that B is a minimal generating set.

"(iii) $\Rightarrow$ (i)": Let B be a minimal generating set. To show that B is a basis, it remains to show that B is linearly independent. Suppose not. Find a nontrivial relation $a_1 v_1 + \cdots + a_n v_n = 0$ for v_1, v_2, ..., $v_n \in B$ and without loss of generality we may assume that a_1, a_2, ..., $a_n \in F$ are all nonzero. This implies that

$$v_1 = a_1^{-1}(-a_2 v_2 - \cdots - a_n v_n) \in \mathrm{Sp}\{v_2, \ldots, v_n\} \subseteq \mathrm{Sp}(B \setminus \{v_1\}).$$

We have that $\mathrm{Sp}(B \setminus \{v_1\}) = \mathrm{Sp}\, B = V$ by Lemma 1.2.5(c). This contradicts to the fact that B is a minimal generating set. Hence B is a linearly independent subset and a basis. $\qquad \square$

Definition 1.3.14. Let V be an F-vector space where F is a field. We say that V is a **finite dimensional** vector space over F if it has a finite generating set for V over F. Otherwise, we say it is an **infinite dimensional** vector space over F.

Every vector space has a basis

One of the most important and fundamental results regarding vector spaces is that every vector space has a basis. In other words, every vector space is free! This result is what separates vector spaces from modules.

Proposition 1.3.15. *Any finite dimensional vector space contains a* finite *basis. In fact, any finite generating set may be reduced to a basis.*

Proof. Find a *finite* generating set for the given finite dimensional vector space. If this generating set is not minimal, we may find a proper generating subset within this generating set. Continue with the procedure and we may eventually reach a minimal generating subset since we started with a *finite* set. From Proposition 1.3.13 this minimal generating subset is a basis for the given vector space. $\qquad \square$

Suppose given an infinite dimensional vector space. Any generating set is infinite. There is no guarantee that we may reduce an *infinite* generating set to a minimal generating subset. To determine whether an infinite dimensional vector space contains a basis, there still remains much to be done.

Definition 1.3.16. We say a poset $(S, \leq)$ is a **totally ordered set** or a **chain** if for all $a, b \in S$ we have either $a \leq b$ or $b \leq a$.

In a totally ordered set any two elements are comparable.

Example 1.3.17. Let $\leq$ be the relation "less than or equal to" in the usual sense. The posets $(\mathbb{Z}_+, \leq)$ and $(\mathbb{Z}, \leq)$ are totally ordered.

Zorn's Lemma. *If every chain in a nonempty poset $(S, \leq)$ has an upper bound in S, S has a maximal element.*

Remark. Any element in S is automatically an upper bound for the empty chain. There is no need to check the empty chain when we verify Zorn's lemma.

This is a well-known *axiom* in set theory. An axiom is a rule we assume in a logical system. It is a statement or a proposition which cannot be proven right or wrong using only logics (and other axioms). In a different logical universe, an axiom may well be assumed to be false. Mathematicians prefer to prove theorems using as few axioms as possible. Nonetheless, Zorn's lemma is one of the most commonly used axioms. Without it, much of what we know in mathematics would be un-established. It would be extremely inconvenient indeed. Zorn's lemma is how human beings imagine infinity, or beyond infinity, is like. When dealing with infinitely countable sets we might be able to use induction. When we have to deal with an uncountable set, Zorn's Lemma is often the only tool we have.

We will now use Zorn's lemma to prove the main result of this section for infinite dimensional vector spaces.

Theorem 1.3.18. *Let F be a field and let V be an F-vector space.*

(a) *Any linearly independent subset of V can be extended to a maximal linearly independent subset.*

(b) *Any generating set of V can be reduced to a minimal generating set.*

Hence, any vector space contains a basis.

Proof. (a) Consider $\mathscr{F}$ which is the family of linearly independent subsets within V. This is a nonempty family since it contains at least the empty set. Clearly $(\mathscr{F}, \subseteq)$ is a poset.

There is nothing to check for the empty chain. We will assume that $\mathscr{C}$ is a nonempty chain in $\mathscr{F}$. We want to find an upper bound for $\mathscr{C}$ in $\mathscr{F}$.

Claim. The set $T = \bigcup_{S \in \mathscr{C}} S$ is a linearly independent subset. Hence $T \in \mathfrak{F}$.

Proof of claim. Suppose T is not linearly independent. This says that we may find $v_1, v_2, \ldots, v_n \in T$ and nonzero elements $a_1, a_2, \ldots, a_n \in F$ such that

$$(1.3.1) \qquad a_1 v_1 + a_2 v_2 + \cdots + a_n v_n = 0.$$

For each i, find $S_i \in \mathscr{C}$ such that $v_i \in S_i$. However, since $\mathscr{C}$ is a chain, this says that one of the S_i's, say S_{i_0}, must contain all the S_i's. The nontrivial relation in (1.3.1) is also a nontrivial relation in S_{i_0}. This says that S_{i_0} is linearly dependent, a contradiction. ∎

The claim shows that the union of $\mathscr{C}$ is an upper bound for $\mathscr{C}$. By Zorn's lemma, $\mathscr{F}$ contains a maximal element, which is a maximal linearly independent subset B in V. By Proposition 1.3.13, the set B is a basis for V over F.

We leave the proof of (b) as an exercise. See Exercise 1. □

Even though we are sure of the existence of bases, it does not mean that we can actually find them. Good luck finding a specific basis for $\prod_{i=1}^{\infty} \mathbb{R}$ over $\mathbb{R}$. Good luck finding a specific basis for $\mathbb{R}$ over $\mathbb{Q}$.

Exercises 1.3

In the following exercises R denotes a ring and F denotes a field.

1. Prove Theorem 1.3.18(b). (Hint: Let X be a generating set for the vector space V over F and let $\mathscr{F}$ be the family of linearly independent subsets within X. Use Zorn's lemma on the poset $(\mathscr{F}, \subseteq)$ to show that there is a maximal linearly independent subset within X and it is a minimal generating set and a basis for V.)

2. Is $\bigoplus_{i=1}^{\infty} R$ a free module over R? If yes, find an R-basis for it. Is $\bigoplus_{i \in \mathbb{R}} R$ a free module over R? If yes, find an R-basis for it.

3. Let V be the subset

$$\{(x_1, x_2, x_3, x_4, x_5) \in \mathbb{C}^5 : 3x_1 + 2x_2 - x_5 = 0, \ x_1 - 3x_3 + x_4 - 2x_5 = 0\}$$

   in V. Show that V is a subspace of $\mathbb{C}^5$ either over $\mathbb{R}$ or over $\mathbb{C}$. Find a basis for V over $\mathbb{C}$. Find a basis for V over $\mathbb{R}$.

4. Let $R[x, y]$ be the polynomial ring of two variables over R. Is it a free module over R? If yes, find a free R-basis for it.

5. Let $S = \{1, 2, 3\}$. Show that $(R^2)^S$ is free over R by finding an R-basis for it.

6. Let $S = \{a_1, a_2, \ldots, a_m\}$ be a set of m elements and let M be a free R-module with a basis $B = \{f_1, f_2, \ldots, f_n\}$. Show that M^S is free over R by finding an R-basis for it.

7. Let I be a proper ideal of R. Use Zorn's lemma to show that in R there is a maximal ideal which contains I. In particular, any ring has a maximal ideal.

1.4 Dimension for finite dimensional vector spaces

In this and the next section, we aim to show that any two bases of a vector space are of the same size. The approaches to the finite dimensional case and to the infinite dimensional case are totally different. In this section we will focus on the finite dimensional case.

Throughout this section F denotes a field.

Replacement theorem

Theorem 1.4.1. *All bases of a finite dimensional vector space are finite and have the same number of elements in them.*

Theorem 1.4.1 can be proved by using the following result.

Proposition 1.4.2 (Replacement theorem). *Let $S = \{v_1, v_2, \ldots, v_n\}$ be a linearly independent set of n elements in the F-vector space V and let B be an F-basis for V. There exist n distinct elements $w_1, w_2, \ldots, w_n$ in B such that*

$$B_i = (B \setminus \{w_1, w_2, \ldots, w_i\}) \cup \{v_1, v_2, \ldots, v_i\}$$

remains an F-basis for V for each $1 \leq i \leq n$.

In particular, we have $|S| \leq |B|$.

Proof. Without loss of generality, we may assume

$$B \cap S = \{v_1, v_2, \ldots, v_s\}$$

where s may be 0. We may replace v_i by itself for $i = 1, \ldots, s$. That is, we would choose $w_i = v_i$ for $i = 1, \ldots, s$. In this case $B = B_1 = B_2 = \cdots = B_s$. If $s < n$, we will attempt to find

$$w_{s+1} \in B \setminus \{v_1, v_2, \ldots, v_s\} = B \setminus \{w_1, w_2, \ldots, w_s\}$$

such that

$$\begin{aligned} B_{s+1} &= (B_s \setminus \{w_{s+1}\}) \cup \{v_{s+1}\} \\ &= (B \setminus \{w_1, w_2, \ldots, w_{s+1}\}) \cup \{v_1, v_2, \ldots, v_{s+1}\} \end{aligned}$$

remains a basis for V over F.

Remember that B is a maximal linearly independent set in V by Proposition 1.3.13. We have that $B \cup \{v_{s+1}\}$ is linearly dependent since $v_{s+1} \notin B$. In $B \cup \{v_{s+1}\}$ we may find a nontrivial relation

$$(1.4.1) \quad av_{s+1} + a_1 v_1 + a_2 v_2 + \cdots + a_s v_s + b_1 u_1 + b_2 u_2 + \cdots + b_d u_d = 0$$

where $a, a_1, a_2, \ldots, a_s, b_1, b_2, \ldots, b_d$ are scalars in F and $u_1, u_2, \ldots, u_d$ are vectors in $B \setminus \{w_1, w_2, \ldots, w_s\}$. First note that if $a = 0$, this will become a nontrivial relation in B_s, a contradiction to the linear independency of B_s. Hence $a \neq 0$. Next, note that $d > 0$, for otherwise we have a nontrivial relation in $\{v_1, v_2, \ldots, v_{s+1}\} \subseteq S$, a contradiction to the linear independency of S. We may also assume that $b_1, b_2, \ldots, b_d$ are all nonzero scalars. Choose $w_{s+1} = u_1$. The choice of w_{s+1} guarantees that $w_1, w_2, \ldots, w_{s+1}$ are distinct elements in B.

Next we will show that B_{s+1} is a basis for V over F. Note that w_{s+1} is the only element in B_s which is not in B_{s+1}. From (1.4.1) we have

$$w_{s+1} = -b_1^{-1}(av_{s+1} + a_1v_1 + a_2v_2 + \dots$$
$$+ a_sv_s + b_2u_2 + b_3u_3 + \dots + b_du_d) \in \mathrm{Sp}\, B_{s+1}.$$

This implies that $B_s \subseteq \mathrm{Sp}\, B_{s+1}$. Thus $V = \mathrm{Sp}\, B_s \subseteq \mathrm{Sp}\, B_{s+1}$. We have that B_{s+1} spans V over F.

Next we check that B_{s+1} is linearly independent over F. Let

$$\alpha_1, \alpha_2, \dots, \alpha_{s+1}, \beta_1, \beta_2, \dots, \beta_k \in F$$

and

$$x_1, x_2, \dots, x_k \in B \setminus \{w_1, w_2, \dots, w_{s+1}\} = B \setminus \{v_1, v_2, \dots, v_s, u_1\}$$

be such that

$$\alpha_1v_1 + \alpha_2v_2 + \dots + \alpha_{s+1}v_{s+1} + \beta_1x_1 + \beta_2x_2 + \dots + \beta_kx_k = 0.$$

Using (1.4.1) this relation may be rewritten as

$$0 = \alpha_1v_1 + \alpha_2v_2 + \dots + \alpha_sv_s$$
$$+ \alpha_{s+1}\big[-a^{-1}(a_1v_1 + \dots + a_sv_s + b_1u_1 + b_2u_2 + \dots + b_du_d)\big]$$
$$+ \beta_1x_1 + \beta_2x_2 + \dots + \beta_kx_k$$
$$= -\alpha_{s+1}a^{-1}b_1u_1$$
$$+ \text{a linear combination of } v_1, v_2, \dots, v_s, u_2, \dots, u_d, x_1, x_2, \dots, x_k$$
$$= -\alpha_{s+1}a^{-1}b_1w_{s+1} + \text{a linear combination of elements in } B_s \setminus \{w_{s+1}\}.$$

Remember that $w_{s+1} \in B_s$. We have that

$$\alpha_{s+1}a^{-1}b_1 = 0 \implies \alpha_{s+1} = 0 \quad \text{since } a^{-1}b_1 \neq 0$$
$$\implies \alpha_1v_1 + \alpha_2v_2 + \dots + \alpha_sv_s + \beta_1x_1 + \beta_2x_2 + \dots + \beta_kx_k = 0$$
$$\implies \alpha_1 = \alpha_2 = \dots = \alpha_s = \beta_1 = \beta_2 = \dots = \beta_k = 0$$

from the linear independence of B_s. Thus B_{s+1} is linearly independent over F. We conclude that B_{s+1} is indeed an F-basis for V.

Note that B_{s+1} and S have $s+1$ elements in common. If $s+1 < n$, we may repeat this process with B_{s+1} and v_{s+2}. Eventually we will exhaust all elements in S. □

Proof of Theorem 1.4.1. Find a finite generating set for the finite dimensional vector space V. This finite generating set may be reduced to a finite basis B by Proposition 1.3.15. Let B' be any other basis for V. From Proposition 1.4.2, we have that $|B| \leq |B'|$ since B is a *finite* linearly independent set. If $|B'| > |B|$, find a subset S of $|B| + 1$ elements in B'. Being a subset of a linearly independent set, S is also linearly independent from Lemma 1.2.13(a). From Proposition 1.4.2 again, $|S| = |B| + 1 \leq |B|$, a contradiction. We conclude that B' is finite and $|B'| = |B|$.   □

Corollary 1.4.3. *An F-vector space is infinite dimensional if and only if it contains an infinite linearly independent subset.*

Proof. "Only if": Let V be an infinite dimensional F-vector space. Then V must be nontrivial. Find $v_1 \neq 0$ in V. Then $\{v_1\}$ is linearly independent over F by Proposition 1.2.18(a). Note that $V \neq \mathrm{Sp}\{v_1\}$ for that would make V finite dimensional. Hence we may find $v_2 \notin \mathrm{Sp}\{v_1\}$. From Proposition 1.2.18(b) we have that $\{v_1, v_2\}$ is linearly independent over F. Similarly, $V \notin \mathrm{Sp}\{v_1, v_2\}$, and we may find $v_3 \notin \mathrm{Sp}\{v_1, v_2\}$ to construct a linearly independent subset $\{v_1, v_2, v_3\}$. Proceed as such we may construct a linearly independent subset $S_n = \{v_1, v_2, \ldots, v_n\}$ for all n. We claim that $S = \{v_1, v_2, \ldots, v_n, \ldots\}$ is an infinite linearly independent subset in V. Let $T = \{v_{i_1}, v_{i_2}, \ldots, v_{i_n}\}$ be any finite subset of S. Without loss of generality we may assume that $i_1 < i_2 < \cdots < i_n$. Being a subset of the linearly independent subset S_{i_n}, T is linearly independent by Lemma 1.2.13(a). Thus S is linearly independent by Definition 1.2.12.

"If": If V is a finite dimensional vector space, we may find a finite basis B of n elements for V by Theorem 1.4.1. If S is a linearly independent subset of more than n elements in V, let T be a subset of $n + 1$ elements in S. Then T is linearly independent over F by Lemma 1.2.13(a). From Proposition 1.4.2, we have that $n + 1 < n$, a contradiction. Thus $|S| \leq n$ is finite.   □

Remark. To prove Theorem 1.4.1, the replacement theorem and Corollary 1.4.3, we very carefully avoided using results whose proofs require Zorn's lemma (we are referring to Theorem 1.3.18 in particular). Finite dimensional vector spaces are not as intangible as their infinite dimensional counterparts after all.

Dimension and rank

Definition 1.4.4. Let V be a finite dimensional F-vector space. Define the **dimension** of V over F, denoted $\dim_F V$, or simply $\dim V$ when F is understood, to be the number of elements in a basis of V over F. If V is an infinite dimensional vector space over F, we write $\dim_F V = \infty$.

Example 1.4.5. From Example 1.3.5 and Corollary 1.4.3 we have the following results.

(a) The direct product: $\dim_F F^n = n$.

(b) The matrix space: $\dim_F M_{m \times n}(F) = mn$.

(c) Let x be an indeterminate over F. Then

$$\dim_F F[x] = \infty \quad \text{and} \quad \dim_F \mathscr{P}_n = n + 1.$$

(d) If S is a finite set, then $\dim_F F^S = |S|$. If S is infinite, then $\dim_F F^S = \infty$. See Exercise 2, §1.2.

Example 1.4.6. Let V be an n-dimensional vector space over $\mathbb{C}$. We claim that V is also a $2n$-dimensional vector space over $\mathbb{R}$.

Let $A = \{v_1, v_2, \ldots, v_n\}$ be a $\mathbb{C}$-basis of V. We will check that

$$B = \{v_1, v_2, \ldots, v_n, iv_1, iv_2, \ldots, iv_n\}$$

is an $\mathbb{R}$-basis of V. Let $v \in V$. Since A is a $\mathbb{C}$-basis of V, we may find $\alpha_1, \alpha_2, \ldots, \alpha_n \in \mathbb{C}$ such that

$$v = \alpha_1 v_1 + \alpha_2 v_2 + \cdots + \alpha_n v_n.$$

Let $\alpha_k = a_k + b_k i$ where a_k and $b_k \in \mathbb{R}$ for $k = 1, 2, \ldots, n$. Then

$$v = a_1 v_1 + a_2 v_2 + \cdots + a_n v_n + b_1 i v_1 + b_2 i v_2 + \cdots + b_n i v_n.$$

Thus V is generated by B over $\mathbb{R}$. To see that B is linearly independent over $\mathbb{C}$, let

$$a_1 v_1 + a_2 v_2 + \cdots + a_n v_n + b_1 i v_1 + b_2 i v_2 + \cdots + b_n i v_n = 0.$$

Let $\alpha_k = a_k + b_k i$ for $k = 1, 2, \ldots, n$. Then

$$\alpha_1 v_1 + \alpha_2 v_2 + \cdots + \alpha_n v_n = 0.$$

Since v_1, v_2, $\ldots$, v_n are linearly independent over $\mathbb{C}$, we have that $\alpha_k = 0$ for all k. This implies that $a_k = b_k = 0$ for all k. We conclude that B is a basis of V over $\mathbb{R}$.

We used Zorn's lemma to prove Theorem 1.3.18, which is a very useful result. Here we also provide a proof without using Zorn's lemma for its finite dimensional case.

Proposition 1.4.7. *Let V be a finite dimensional F-vector space. The following assertions are true.*

(a) *Any linearly independent subset may be enlarged to a basis.*

(b) *Any generating set may be reduced to a basis.*

Proof. If V is the trivial space, the only linearly independent set in V and the only basis of V is the empty set $\varnothing$! Without loss of generality, we will assume that V is non-trivial.

From Proposition 1.3.15, we know that a finite dimensional vector space V has a finite basis of, say, n elements.

(a) Let S be a linear independent subset of V. First note that $|S| \leq n$ from the Replacement Theorem (Proposition 1.4.2). If $V = \mathrm{Sp}\, S$ then S is a basis for V and we are done. Otherwise, find $v \notin \mathrm{Sp}\, S$. We have that $S \cup \{\, v \,\}$ is linearly independent by Proposition 1.2.18(c). Repeat this process with the set $S \cup \{\, v \,\}$. Eventually we will reach a linearly independent subset which also spans V, for otherwise we will have a linearly independent subset of more than n elements, a contradiction. This linearly independent subset which spans V is a basis for V.

(b) Let S be a generating set for V. Then $S \neq \varnothing$ and $S \neq \{\, 0 \,\}$ since V is nontrivial. Before we start, remember that the only subspace containing S is V.

Find a nonzero vector v_1 in S. If $\mathrm{Sp}\{\, v_1 \,\} \neq V$, then $\mathrm{Sp}\{\, v_1 \,\}$ cannot contain S. In this case we may find $v_2 \in S \setminus \mathrm{Sp}\{\, v_1 \,\}$. If $\mathrm{Sp}\{\, v_1, v_2 \,\} \neq V$, we may find $v_3 \in S \setminus \mathrm{Sp}\{\, v_1, v_2 \,\}$, etc.

If the aforementioned process cannot stop, we will have a sequence of elements v_1, v_2, ..., v_{n+1} in S such that

- $v_1 \neq 0$;

- $v_i \notin \mathrm{Sp}\{ v_1, v_2, \ldots, v_{i-1} \}$ for $i = 2, 3, \ldots, n+1$.

Form Proposition 1.2.18(b), $\{ v_1, v_2, \ldots, v_{n+1} \}$ is a linearly independent subset in V, a contradiction to Proposition 1.4.2. Hence we should have a sequence of elements v_1, v_2, ..., v_t in S such that

- $v_1 \neq 0$;

- $v_i \notin \mathrm{Sp}\{ v_1, v_2, \ldots, v_{i-1} \}$ for $i = 2, 3, \ldots, t$;

- $\mathrm{Sp}\{ v_1, v_2, \ldots, v_t \} = V$.

Form Proposition 1.2.18(b), $\{ v_1, v_2, \ldots, v_t \}$ is linearly independent, and thus a basis for V. (In fact, $t = n$ by Theorem 1.4.1.) $\square$

This result does not hold for free modules as we can see in the following example.

Example 1.4.8. Consider $\mathbb{Z}$ as a $\mathbb{Z}$-module. Both sets $\{ 2, 3 \}$ and $\{ 1 \}$ are minimal generating sets. These two minimal generating sets are of different sizes. The set $\{ 1 \}$ is a free basis for $\mathbb{Z}$ over $\mathbb{Z}$. The set $\{ 2, 3 \}$ cannot be reduced to a free basis for $\mathbb{Z}$ over $\mathbb{Z}$.

Dimension is an important feature for any vector space.

Proposition 1.4.9. *Let V be an n-dimensional F-vector space. The following conditions are equivalent.*

(i) *The set $\{ v_1, v_2, \ldots, v_n \}$ is a basis for V.*

(ii) *The set $\{ v_1, v_2, \ldots, v_n \}$ is linearly independent over F.*

(iii) *The set $\{ v_1, v_2, \ldots, v_n \}$ spans V over F.*

Proof. Let $B = \{ v_1, v_2, \ldots, v_n \}$.
   "(i) $\Rightarrow$ (ii)": This holds by definition.

"(ii) $\Rightarrow$ (iii)": If B is not a basis, the set B may be enlarged to a basis of more than n elements by Proposition 1.4.7(a), a contradiction to Theorem 1.4.1. Hence B is a basis and spans V.

"(iii) $\Rightarrow$ (i)": If B is not a basis, the set B may be reduced to a basis of less than n elements by Proposition 1.4.7(b), a contradiction to Theorem 1.4.1. Thus B is a basis. $\square$

Corollary 1.4.10. *Let V be a finite dimensional vector space over F and let W be a proper subspace of V. Then $\dim W < \dim V$.*

Proof. Let S be a basis for W over F. This is a linearly independent subset in V. By Proposition 1.4.7(b), we may enlarge it to a *finite* basis B for V. Since $\operatorname{Sp} S = W \subsetneq V$, the set B properly contains S. Hence $\dim W = |S| < |B| = \dim V$. $\square$

Suppose M is a free R-module with a finite basis. We say that M is a free module of **finite rank**. It can be shown that any two bases of M are of the same size, but we will leave the proof until later in this book. The size of any basis is called the **rank** or **free rank** of the free module. The concept of *rank* is more tricky than the concept of *dimension*. If R is non-commutative, it may happen that a free module M possesses two bases of different sizes.

Exercises 1.4

Throughout these exercises, F denotes a field.

1. Let V be the vector space in Exercise 3, §1.3. Find $\dim_{\mathbb{C}} V$ and $\dim_{\mathbb{R}} V$.

2. Let S be an m-element set and let V be an n-dimensional vector space over F. Find $\dim_F V^S$.

3. Consider the $\mathbb{Z}_5$-vector space $V = \mathbb{Z}_5^4$.

   (a) Show that $w_1 = (1,1,1,1)$ and $w_2 = (1,2,3,4)$ are linearly independent over $\mathbb{Z}_5$.

(b) Show that

$$f_1 = (1, 3, 0, 1), \ f_2 = (2, 3, 0, 5), \ f_3 = (0, 1, 3, 3), \ f_4 = (0, 1, 1, 3)$$

form a basis for V over $\mathbb{Z}_5$.

(c) Replace elements in the basis $\{ f_1, f_2, f_3, f_4 \}$ by w_1 and w_2 to obtain a new basis for V over $\mathbb{Z}_5$.

4. Let $F[x]$ be the polynomial ring of one indeterminate over F and let $f(x) \in F[x]$. We say $f(x)$ is an **even** polynomial if $f(-x) = f(x)$. We say $f(x)$ is an **odd** polynomial if $f(-x) = -f(x)$.

   Let $\mathscr{E}_n$ be the set of even polynomials of degree $\leq n$ in $F[x]$. Let $\mathscr{O}_n$ be the set of odd polynomials of degree $\leq n$ in $F[x]$. Show that $\mathscr{E}_n$ and $\mathscr{O}_n$ are F-subspaces of $F[x]$. Find $\dim_F \mathscr{E}_n$ and $\dim_F \mathscr{O}_n$.

5. Let V be a finite-dimensional vector space over F. Let W and W' be subspaces of V such that $\dim W = \dim W'$. Show that $W = W'$ if and only if $W \subseteq W'$.

6. Let $F \subseteq E$ be a field extension. Let V be an E-vector space.

   (a) Show that $\dim_F V = mn$ if $\dim_F E = n$ and $\dim_E V = m$.

   (b) Show that $\dim_F V = \infty$ if $\dim_E V = \infty$.

   (c) Show that $\dim_F V = \infty$ if $\dim_F E = \infty$.

7. Let U be a finite-dimensional F-vector space and let V and W be its subspaces. Show that

$$\dim(V \cap W) + \dim(V + W) = \dim V + \dim W.$$

8. Let V and W be finite-dimensional F-vector spaces. S how that

$$\dim(V \oplus W) = \dim V + \dim W.$$

(For the definition of $V \oplus W$ see Exercise 7, §1.1.)

9. Let W be a subspace of the vector space V over F. If U is a subspace of V such that

$$U + W = V \quad \text{and} \quad U \cap W = \{0\},$$

we say that U is a **complement** of W in V.

(a) Let U be a complement of W in V. Suppose $\{u_1, u_2, \ldots, u_k\}$ is an F-basis for U and $\{w_1, w_2, \ldots, w_m\}$ is an F-basis for W. Show that $\{u_1, u_2, \ldots, u_k, w_1, w_2, \ldots, w_m\}$ is an F-basis for V.

(b) Show that any subspace of V has a complement in V.

(c) Let W be a non-trivial and proper subspace of V. Does W have a unique complement in V?

(d) Let $\dim_F V = n$ and $\dim_F W = m$. Show that every complement of W in V is of dimension $n - m$.

1.5 Dimension for infinite dimensional vector spaces

In order to obtain a similar conclusion to that of Theorem 1.4.1 for the infinite dimensional case, we need to know how to compare the sizes of two sets first.

How to measure the size of a set

Definition 1.5.1. We say that X is **equivalent** to Y, denoted $X \sim Y$, if there is a bijective map from X to Y.

Here, $\sim$ is indeed an equivalence relation. See Exercise 1.

It is fair to say that two sets are of the same size if they are equivalent. However, it is often infeasible, or cumbersome to say the least, to establish a bijection between two sets even when we are sure of the fact. Therefore, we need to find a different approach.

Let X and Y be two sets. If the elements of X can be made to correspond with the elements of a subset of Y, it is reasonable to imagine that the size of X is no greater than the size of Y.

Definition 1.5.2. We say that Y **dominates** X, denoted $X \preceq Y$, if there is an injective map from X to Y.

This is just a fancy way to say that Y is a "bigger" set compared with X. For this to work, we first observe that domination is "almost" a partial order. Both reflexivity and transitivity are obvious. Clearly, $X \xrightarrow{1_X} X$ is an injective map. If $X \xrightarrow{f} Y$ and $Y \xrightarrow{g} Z$ are both injective, then $X \xrightarrow{gf} Z$ is also injective. The following result gives the quasi-antisymmetry of $\preceq$.

Theorem 1.5.3 (Schröder–Bernstein theorem). *If $X \preceq Y$ and $Y \preceq X$, then $X \sim Y$.*

Proof. To avoid confusion and without loss of generality, we will assume that X and Y have no elements in common. To achieve this, we may replace X by a copy of X, say $X' \times \{0\}$, and we may replace Y by $Y' = Y \times \{1\}$. There is a natural bijection $\pi_X \colon X \to X'$ sending x to $(x, 0)$. Similarly, there is also a natural bijection $\pi_Y \colon Y \to Y'$ sending y to $(y, 1)$. If we can construct a bijection $\pi \colon X' \to Y'$, we also have a bijection $\pi_Y^{-1} \circ \pi \circ \pi_X$ from X onto Y:

$$ X \xrightarrow[\text{bij.}]{\pi_X} X' \xrightarrow[\text{bij.}]{\pi} Y' \xrightarrow[\text{bij.}]{\pi_Y^{-1}} Y. $$

Let $X \xrightarrow{f} Y$ and $Y \xrightarrow{g} X$ both be injective maps. We say that an element x in X is the *parent* of the element $f(x)$ in Y, and similarly, an element $y \in Y$ is the parent of the element $g(y)$ in X.

If we trace the ancestry of an element as far back as possible, we

- either ultimately arrive at an element with no parent (the orphans are exactly the elements in $X - g(Y)$ or in $Y - f(X)$),

- or find the lineage continues ad infinitum.

Let X_X be the set of elements in X which originate in X, X_Y be the set of elements in X which originate in Y, and X_∞ be the set of elements in X which have no parentless ancestors. Define Y_X, Y_Y and Y_∞ similarly. Then $X = X_X \,\dot\cup\, X_Y \,\dot\cup\, X_\infty$ and $Y = Y_X \,\dot\cup\, Y_Y \,\dot\cup\, Y_\infty$. Here the notation $\dot\cup$, the **disjoint union**, indicates that the union is formed without the components overlapping. Clearly, among all these sets X_X, X_Y, X_∞, Y_X, Y_Y and Y_∞, none of them intersect with each other.

Next we will show that $X_X \sim Y_X$, $Y_Y \sim X_Y$ and $X_\infty \sim Y_\infty$. Define $f_X \colon X_X \to Y_X$ to be the function sending x to $f(x)$. Clearly f_X is injective. Let $y \in Y_X$. Since y has an ancestor in X, it must have a parent x in X.

Thus $f_X(x) = y$. This shows that f_X is onto as well. Hence f_X is a bijection and $X_X \sim Y_X$.

Define $g_y \colon Y_Y \to X_Y$ to be the function sending y to $g(y)$. One can argue similarly that g_Y is an bijection and $Y_Y \sim X_Y$.

Finally, define $f_\infty \colon X_\infty \to Y_\infty$ to be the function sending x to $f(x)$. Similarly f_∞ is injective. Let $y \in Y_\infty$. It must have a parent, say x, in X. Hence $f_\infty(x) = y$. This shows that f_∞ is onto. Thus f_∞ is an bijection and $X_\infty \sim Y_\infty$.

We now can construct the function

$$
\begin{array}{rccc}
\Phi \colon & X & \longrightarrow & Y \\
& x \in X_X & \longmapsto & f(x) \\
& x \in X_Y & \longmapsto & g_Y^{-1}(y) \\
& x \in X_\infty & \longmapsto & f(x).
\end{array}
$$

The function Φ is an bijection and $X \sim Y$. $\square$

Definition 1.5.4. We write $|X| \leq |Y|$ if $X \preceq Y$, and we say two sets X and Y are of the same **cardinality**, denoted $|X| = |Y|$, if $X \sim Y$.

Since $\sim$ is an equivalence relation, we may think of $|X|$ as the equivalence class of X with respect to $\sim$. Schröder–Bernstein theorem gives the antisymmetry of $\leq$. We now have that $\leq$ is a partial order on the cardinality of sets.

To show two sets A and B are of the same size, instead of constructing a bijective map between them, we may choose to construct two injective maps $A \hookrightarrow B$ and $B \hookrightarrow A$.

A brief discussion on cardinal numbers

Now we will talk briefly about how mathematicians envision the size of a set. The discussion in this subsection will be given without proof. For a thorough treatment please refer to a standard textbook on set theory.

The axiom of choice. Let $\{X_i\}_{i \in I}$ be a nonempty family of nonempty sets. Then $\prod_{i \in I} X_i \neq \varnothing$.

An element $(x_i)_{i \in I}$ in $\prod_{i \in I} X_i$ gives a function f whose domain is I and $f(i) \in X_i$ for $i \in I$. The function f is called a *choice* function.

Definition 1.5.5. A poset $(S, \leq)$ is **well-ordered** if every nonempty subset of S has a least element.

Well-ordering principle. Every set can be well-ordered.

For someone who is naïve in set theory, the axiom of choice seems trivial while the well-ordering principle seems to be false or incomprehensible. These two principles are in fact *axioms*, and they are both equivalent to Zorn's lemma. They are powerful tools for treating infinite sets. However, let's not be greedy right away and let's start by reacquainting ourselves with old friends.

The natural numbers may be constructed as follows.

Let $0 = \varnothing$. Once a natural number n is constructed, we may construct a *successor* to n, denoted as n^+, by letting $n^+ = n \cup \{n\}$. Thus

$$0 = \varnothing;$$
$$1 = 1^+ = 0 \cup \{0\} = \{0\};$$
$$2 = 1^+ = 1 \cup \{1\} = \{0, 1\}$$
$$3 = 2^+ = 2 \cup \{2\} = \{0, 1, 2\}$$
$$4 = 3^+ = 3 \cup \{3\} = \{0, 1, 2, 3\}$$

etc.

Define $\omega = \{0, 1, 2, 3, \dots\}$ to be the set of *natural numbers*.

There is no need to stop at ω. We may continue to construct the successor to ω:

$$\omega^+ = \omega \cup \{\omega\}.$$

For some obvious reason people refer to ω^+ as $\omega + 1$ and $(\omega^+)^+$ as $\omega + 2$, etc. Thus an infinite sequence of "numbers" are born:

$$\omega, \ \omega + 1, \ \omega + 2, \ \omega + 3, \ \dots.$$

We may lump all these numbers together into one set and call it $\omega + \omega$, or more often, people call it 2ω. We may continue to add successors from numbers already constructed, and when the act of adding successors is exhausted we may lump all constructed numbers together to form a new gigantic set (number). And then we have a new number to add new successors to again. This process can go on and on forever:

$$\cdots, \ 2\omega, \ 2\omega + 1, \ 2\omega + 2, \ \ldots, \ 3\omega, \ \ldots, \ 4\omega, \ldots,$$
$$\omega\omega = \omega^2, \ \ldots, \ \omega^3, \ \ldots, \ \omega^4, \ \ldots, \ \omega^\omega, \ \ldots.$$

In this fashion, it is possible to construct a *well-ordered* family (with $\subseteq$ as the partial order) which contains ω and sets of any cardinality. We will not go into more details for this is not a book on set theory after all. Every element in this family has a successor but not necessarily a "predecessor". For example, ω does not have a predecessor. The elements in this family are called **ordinal numbers.**

The family of ordinal numbers satisfies the well-ordering principle. Thus the sub-family of ordinal numbers of the same cardinality contains a least element which is called a **cardinal number.** A cardinal number can be used to represent the cardinality of all sets equivalent to itself.

Each natural number n is a *finite* cardinal number. If I have n pebbles in my hand, it means that there is a bijection between the pebbles in my hand and the elements in the set n. Since ω is the first infinite set we encounter in this family, it is a cardinal number.

Let X be an infinite set. We can construct an injection from ω into X as follows. First, well order X. Map 0 to the least element x_0 in S. Map 1 to the least element x_1 in $X \setminus \{x_0\}$. Map 2 to the least element in $X \setminus \{x_0, x_1\}$ and so on. Since X is infinite, $X \setminus \{x_0, x_1, \ldots, x_n\}$ will never be empty for any $n \in \omega$. We may thus map $n + 1$ to the least element x_{n+1} in $X \setminus \{x_0, x_1, \ldots, x_n\}$. This shows that $\omega \preceq X$. In other words, ω is the *least* infinite cardinal number.[3]

If $X \sim \omega$, we write $|X| = \omega$.[4] In this case, we say that X is a **countable** set, or that X is **countably infinite.** If X is an infinite set such that

---

[3]In fact, in order for any infinite set to dominate ω, one only needs to assume the *axiom of countable choice,* denoted $\mathbf{AC}_\omega$.

AC_ω is a weaker form of the *axiom of choice.* It states that every countable family of nonempty sets must have a choice function. More specifically, let $\{X_i\}_{i \in \omega}$ be a countable collection of nonempty sets. There is a function f with domain ω such that $f(n) \in X_n$ for all $n \in \omega$.

[4]The countable cardinal number of ω is also denoted as $\aleph_0$ due to Georg Cantor, 1845–1918.

$|X| \neq \omega$, we say that X is **uncountable**.

Cardinal arithmetic

Now we give a few rules on the arithmetic of cardinality. We will assume that the reader is familiar with how to add or multiply natural numbers together. Our focus will be on the infinite cardinal numbers.

Let X and Y be two sets. We will use $X \mathbin{\dot\cup} Y$ to denote the **disjoint union** of X and Y. This means that we will simply differentiate elements of X from elements of Y even when some of them are the same elements in reality. For example, if $X = \{0, 1\}$ and $Y = \{1, 2\}$, to create the disjoint union of X and Y, we replace the 1 in Y by a clone, say $1'$. Hence, $X \mathbin{\dot\cup} Y = \{0, 1, 1', 2\}$ contains 4 elements. More formally, we may replace every element $x \in X$ by a clone $(x, 0) \in X \times \{0\}$ and every element $y \in Y$ by a clone $(y, 1) \in Y \times \{1\}$. Thus

$$X \mathbin{\dot\cup} Y \overset{\text{def}}{=} \left(X \times \{0\}\right) \cup \left(Y \times \{1\}\right).$$

When $X \cap Y = \varnothing$, we also denote the union by $X \mathbin{\dot\cup} Y$ to emphasize that the two subsets are disjoint. This usually causes no confusion, especially in regard to the cardinality. It is easy to see that in this case $X \cup Y \sim X \mathbin{\dot\cup} Y$.

Definition 1.5.6. For the two sets X and Y, we define $|X| + |Y| = |X \mathbin{\dot\cup} Y|$ and $|X||Y| = |X \times Y|$.

We need to verify that this definition makes sense. We leave the details to the reader. See Exercise 2.

Lemma 1.5.7. *Let X be any set. Then $|X| + 0 = |X|$ and $0|X| = 0$.*[5]

Proof. This follows from the facts that $X \mathbin{\dot\cup} \varnothing \sim X$ and $\varnothing \times X = \varnothing$. $\square$

The following proposition shows that cardinal arithmetic has the same basic properties of the elementary school arithmetic on natural numbers.

Proposition 1.5.8. *Let X, Y, Z and W be sets. The following properties are satisfied:*

---

[5]This statement is true only in the context of set theory. When you see "zero times infinity" elsewhere, it may mean something else.

(a) *Associativity:*

   (i) $(|X| + |Y|) + |Z| = |X| + (|Y| + |Z|);$

   (ii) $(|X||Y|)|Z| = |X|(|Y||Z|).$

(b) *Commutativity:*

   (i) $|X| + |Y| = |Y| + |X|;$

   (ii) $|X||Y| = |Y||X|.$

(c) *Distributivity:* $(|X| + |Y|)|Z| = |X||Z| + |Y||Z|.$

(d) *Inequality: if* $|X| \leq |Z|$ *and* $|Y| \leq |W|,$ *then*

   (i) $|X| + |Y| \leq |Z| + |W|;$

   (ii) $|X||Y| \leq |Z||W|.$

Proof. We will provide a mapping for each case, and we will leave the details to the reader (see Exercise 3).

   (a) The mapping

$$
\begin{array}{rcl}
(X \mathbin{\dot\cup} Y) \mathbin{\dot\cup} Z & \longrightarrow & X \mathbin{\dot\cup} (Y \mathbin{\dot\cup} Z) \\
((x,0),\,0),\ x \in X & \longmapsto & (x,\,0) \\
((y,1),\,0),\ y \in Y & \longmapsto & ((y,0),\,1) \\
(z,\,1),\ z \in Z & \longmapsto & ((z,1),\,1)
\end{array}
$$

and the mapping

$$
\begin{array}{rcl}
(X \times Y) \times Z & \longrightarrow & X \times (Y \times Z) \\
((x,y),\,z) & \longmapsto & (x,\,(y,z))
\end{array}
$$

are bijections.

   (b) The mapping

$$
\begin{array}{rcl}
X \mathbin{\dot\cup} Y & \longrightarrow & Y \mathbin{\dot\cup} X \\
(x,\,0),\ x \in X & \longmapsto & (x,\,1) \\
(y,\,1),\ y \in Y & \longmapsto & (y,\,0)
\end{array}
$$

and the mapping

$$
\begin{array}{rcl}
X \times Y & \longrightarrow & Y \times X \\
(x,\,y) & \longmapsto & (y,\,x)
\end{array}
$$

are bijections.

(c) The mapping

$$(X \mathbin{\dot\cup} Y) \times Z \longrightarrow (X \times Z) \mathbin{\dot\cup} (Y \times Z)$$
$$((x,0),\, z)),\; x \in X,\, z \in Z \longmapsto ((x,z),\, 0)$$
$$((y,1),\, z)),\; y \in Y,\, z \in Z \longmapsto ((y,z),\, 1)$$

is an bijection.

(d) Suppose given injections $\varphi \colon X \to Z$ and $\psi \colon Y \to W$. We define

$$\varphi \mathbin{\dot\cup} \psi \colon\; X \mathbin{\dot\cup} Y \longrightarrow Z \mathbin{\dot\cup} W$$
$$(x,0) \longmapsto (\varphi(x),0)$$
$$(y,1) \longmapsto (\psi(y),1)$$

and

$$\varphi \times \psi \colon\; X \times Y \longrightarrow Z \times W$$
$$(x,y) \longmapsto (\varphi(x),\psi(w)).$$

These two mappings are injections and so we have $|X| + |Y| \le |Z| + |W|$ and $|X||Y| \le |Z||W|$. $\qquad\square$

Lemma 1.5.9. *Let n be a nonzero natural number. The following assertions are true:*

(a) $n + \omega = \omega$ *and* $\omega + \omega = \omega$;

(b) $n\omega = \omega$ *and* $\omega\omega = \omega$.

Proof. We will provide a mapping for each case, and we will leave the details to the reader (see Exercise 4).

(a) The mapping

$$n \mathbin{\dot\cup} \omega \longrightarrow \omega$$
$$(k,\, 0),\; 0 \le k \le n-1 \longmapsto k$$
$$(k,\, 1),\; k \in \omega \longmapsto n+k$$

and the mapping

$$\omega \mathbin{\dot\cup} \omega \longrightarrow \omega$$
$$(k,\, 0) \longmapsto 2k$$
$$(k,\, 1) \longmapsto 2k+1$$

are both bijections.

(b) The mapping

$$n \times \omega \longrightarrow \omega$$
$$(k, \ell) \longmapsto n\ell + k$$

is a bijection. To show that $\omega \sim \omega \times \omega$ we define a lexicographical order on $\omega \times \omega$ be letting $(i, j) \leq (k, \ell)$ if $i + j < k + \ell$ or if $i + j = k + \ell$ together with $i \leq k$. This will give a linear order to $\omega \times \omega$:

$$(0,0) < (0,1) < (1,0) < (0,2) < (1,1) < (2,0) < (0,3) < (1,2) < \cdots$$

There is an bijection from ω to $\omega \times \omega$ by sending 0 to the first element in this sequence, 1 to the second element, 2 to the third, etc. $\square$

For the next result, the main difficulty arises when we have to deal with uncountable sets. Now there is no escape from using Zorn's lemma / Axiom of choice / Well-ordering principle.

Theorem 1.5.10. *If $0 < |X| \leq |Y|$ and Y is an infinite set, the following assertions are true:*

(a) $|X| + |Y| = |Y|$;

(b) $|X||Y| = |Y|$.

Proof. (a) We will prove $|X| + |Y| = |Y|$ in three steps.

Step 1. We prove the case where $|X|$ is finite or countable.

Since Y is infinite, there is a countable subset A within Y. We may write $Y = A \mathbin{\dot\cup} (Y \setminus A)$. In other words, $|Y| = \omega + |Y \setminus A|$. When X is finite or countable, we have $|X| + \omega = \omega$ by Lemma 1.5.9(a). Thus we have that

$$|X| + |Y| = |X| + (\omega + |Y \setminus A|) = (|X| + \omega) + |Y \setminus A|$$
$$= \omega + |Y \setminus A| = |Y|$$

by Proposition 1.5.8(a).

Step 2. We prove that $|Y| = |Y| + |Y|$. Consider the poset $(\mathfrak{S}, \leq)$ where

$$\mathfrak{S} = \{(Z, f) : Z \subseteq Y, \ Z \xrightarrow{f} Z \mathbin{\dot\cup} Z \text{ is a bijection}\}.$$

The partial order is given by letting

$$(Z, f) \leq (Z', f') \text{ if } Z \subseteq Z' \text{ and } f'(z) = f(z) \text{ for all } z \in Z.$$

The infinite set Y must contain a countable subset W. We may find a bijection $h\colon W \to W \mathbin{\dot\cup} W$ by Lemma 1.5.9(a). Thus $(W, h) \in \mathfrak{S}$ and $\mathfrak{S}$ is nonempty.

Let $\mathfrak{C}$ be a nonempty chain in $\mathfrak{S}$. Let $U = \bigcup_{(Z,f) \in \mathfrak{C}} Z$. We want to define a function from U to $U \mathbin{\dot\cup} U$ in the following manner. Let $x \in U$. Find $Z \subseteq Y$ such that $x \in Z$ and $(Z, f) \in \mathfrak{C}$. Define $g(x) = f(x)$. Our first question is whether g is well-defined. Suppose $x \in Z$ and $x \in Z'$ such that (Z, f) and $(Z', f') \in \mathfrak{S}$. Since $\mathfrak{C}$ is a chain, we may assume without loss of generality that $(Z, f) \leq (Z', f')$. Thus $f'(x) = f(x)$ from the definition of $\leq$. This shows that g is well-defined. We proceed to show that g is bijective.

To show that g is injective, suppose $g(x) = g(x')$. Find $(Z, f) \in \mathfrak{C}$ such that $x \in Z$. Find $(Z', f') \in \mathfrak{C}$ such that $x' \in Z'$. We may assume without loss of generality that $(Z, f) \leq (Z', f')$. Then $x \in Z'$ as well. Hence $f'(x) = g(x) = g(x') = f'(x')$. We have that $x = x'$ since f' is injective.

To check that g is surjective, remember that

$$U \mathbin{\dot\cup} U = \{(x, 0) : x \in U\} \cup \{(x, 1) : x \in U\}.$$

For any $x \in U$, find $(Z, f) \in \mathfrak{C}$ such that $x \in Z$. Since f maps Z onto $Z \mathbin{\dot\cup} Z$, we may find z_0 and $z_1 \in Z$ such that $f(z_0) = (x, 0)$ and $f(z_1) = (x, 1)$. Thus $g(z_0) = f(z_0) = (x, 0)$ and $g(z_1) = f(z_1) = (x, 1)$. This implies that g is also surjective.

We have just shown that $(U, g) \in \mathfrak{S}$ and (U, g) is clearly an upper bound for $\mathfrak{C}$. By Zorn's lemma, there is a maximal element $(\widetilde{Y}, \tilde{f})$ in $\mathfrak{S}$. We claim that $Y \setminus \widetilde{Y}$ is finite.

Assume $Y \setminus \widetilde{Y}$ is infinite. We may find $W \subseteq (Y \setminus \widetilde{Y})$ such that W is countable. We can find a bijection $h\colon W \to W \mathbin{\dot\cup} W$ by Lemma 1.5.9(a). Note that

$$(\widetilde{Y} \cup W) \mathbin{\dot\cup} (\widetilde{Y} \cup W)$$
$$= \{(y, 0) : y \in \widetilde{Y}\} \cup \{(w, 0) : w \in W\} \cup \{(y, 1) : y \in \widetilde{Y}\} \cup \{(w, 1) : w \in W\}$$
$$= (\widetilde{Y} \mathbin{\dot\cup} \widetilde{Y}) \cup (W \mathbin{\dot\cup} W).$$

Since $\widetilde{Y} \cap W = \varnothing$, the three sets $\widetilde{Y} \cup W$, $\widetilde{Y} \mathbin{\dot\cup} \widetilde{Y}$ and $W \mathbin{\dot\cup} W$ are all disjoint.

Construct the map

$$\tilde{f} \cup h : \quad \widetilde{Y} \cup W \quad \longrightarrow \quad (Y \,\dot\cup\, Y) \cup (W \,\dot\cup\, W)$$
$$y \in \widetilde{Y} \quad \longmapsto \quad \tilde{f}(y)$$
$$w \in W \quad \longmapsto \quad h(w).$$

We leave it to the reader as an easy exercise to verify that $\tilde{f} \cup h$ is a bijection and that with respect to our order, $(\widetilde{Y} \cup W, \tilde{f} \cup h)$ is strictly greater than $(\widetilde{Y}, \tilde{f})$. This is a contradiction to the maximality of $(\widetilde{Y}, \tilde{f})$.

We now have that $Y \backslash \widetilde{Y}$ is finite. This show that $|Y| = |Y \backslash \widetilde{Y}| + |\widetilde{Y}| = |\widetilde{Y}|$ by Step 1. Hence

$$|Y| = |\widetilde{Y}| = |\widetilde{Y}| + |\widetilde{Y}| = |Y| + |Y|.$$

Step 3. In general, $|Y| = 0 + |Y| \leq |X| + |Y| \leq |Y| + |Y| = |Y|$ by Lemma 1.5.7, Proposition 1.5.8(d) and Step 2. Hence $|X| + |Y| = |Y|$.

(b) Now we will show that $|X||Y| = |Y|$ in two steps.

Step 1. We first prove the case that $|Y||Y| = |Y|$. Consider the poset $(\mathfrak{T}, \leq)$ where

$$\mathfrak{T} = \{(Z, f) : Z \xrightarrow{f} Z \times Z \text{ is bijective}\}$$

and the partial order is given by letting $(Z, f) \leq (Z', f')$ if $Z \subseteq Z'$ and $f'(z) = f(z)$ for all $z \in Z$. Let W be a countable subset within Y. We may find a bijection $h \colon W \to W \times W$ by Lemma 1.5.9(b). Thus $(W, h) \in \mathfrak{T}$ and $\mathfrak{T}$ is nonempty.

Let $\mathfrak{C}$ be a nonempty chain in $\mathfrak{T}$. Let $U = \bigcup_{(Z,f) \in \mathfrak{C}} Z$. We want to define a function from U to $U \times U$ in the following manner. Let $x \in U$. Find $Z \subseteq Y$ such that $u \in Z$ and $(Z, f) \in \mathfrak{C}$. Define $g(x) = f(x)$. Our first question is whether g is well-defined. Suppose $x \in Z$ and $x \in Z'$ such that (Z, f) and $(Z', f') \in \mathfrak{T}$. Since $\mathfrak{C}$ is a chain, we may assume without loss of generality that $(Z, f) \leq (Z', f')$. We have $f'(x) = f(x)$ from the definition of $\leq$. Hence that g is well-defined. We proceed to check that g is bijective.

To check that g is injective, suppose $g(x) = g(x')$. Find $(Z, f) \in \mathfrak{C}$ such that $x \in Z$. Find $(Z', f') \in \mathfrak{C}$ such that $x' \in Z'$. We may assume without loss of generality that $(Z, f) \leq (Z', f')$. Then $x \in Z'$ as well. Hence $f'(x) = g(x) = g(x') = f'(x')$. We have that $x = x'$ since f' is injective.

To check that g is surjective, let $(x, \ x') \in U \times U$. Find (Z, f) and $(Z', f') \in \mathfrak{C}$ such that $x \in Z$ and $x' \in Z'$. If $(Z', f') \leq (Z, f)$, then x, $x' \in Z$ and $(x, \ x') \in Z \times Z$. Since f maps Z onto $Z \times Z$, we may find $z \in Z \subseteq U$ such that $g(z) = f(z) = (x, \ x')$. Similarly, if $(Z, f) \leq (Z', f')$, then $(x, \ x') \in Z' \times Z'$. Since f' maps Z' onto $Z' \times Z'$, we may find $z' \in Z' \subseteq U$ such that $g(z') = f'(z') = (x, \ x')$. This shows that g is surjective.

We have just shown that $(U, g) \in \mathfrak{T}$ and (U, g) is clearly an upper bound for $\mathfrak{C}$. By Zorn's lemma again, there is a maximal element $(\widetilde{Y}, \tilde{f})$ in $\mathfrak{T}$. We have that $|\widetilde{Y}||\widetilde{Y}| = |\widetilde{Y}|$. Next we show that $|Y \setminus \widetilde{Y}| < |Y|$.

Suppose that $|Y \setminus \widetilde{Y}| = |Y|$. Establish a bijection from Y onto $Y \setminus \widetilde{Y}$. Then the image of $\widetilde{Y}$ within $Y \setminus \widetilde{Y}$, denoted $Y^\sharp$, is such that $|Y^\sharp| = |\widetilde{Y}|$. The choice of $Y^\sharp$ guarantees that $|Y^\sharp| = |Y^\sharp||Y^\sharp|$. Since $\widetilde{Y}$ and $Y^\sharp$ are disjoint, we know that the three sets $\widetilde{Y} \times Y^\sharp$, $Y^\sharp \times \widetilde{Y}$ and $Y^\sharp \times Y^\sharp$ are disjoint from each other. We have that

$$
\begin{aligned}
&|(\widetilde{Y} \times Y^\sharp) \mathbin{\dot\cup} (Y^\sharp \times \widetilde{Y}) \mathbin{\dot\cup} (Y^\sharp \times Y^\sharp)| \\
&= |\widetilde{Y} \times Y^\sharp| + |Y^\sharp \times \widetilde{Y}| + |Y^\sharp \times Y^\sharp| \\
&= |\widetilde{Y}||Y^\sharp| + |Y^\sharp||\widetilde{Y}| + |Y^\sharp||Y^\sharp| \\
&= |Y^\sharp||Y^\sharp| + |Y^\sharp||Y^\sharp| + |Y^\sharp||Y^\sharp| \\
&= |Y^\sharp| + |Y^\sharp| + |Y^\sharp| \\
&= |Y^\sharp|, \qquad \text{by (a)}.
\end{aligned}
$$

Thus we can find a bijection

$$
h \colon Y^\sharp \to (\widetilde{Y} \times Y^\sharp) \mathbin{\dot\cup} (Y^\sharp \times \widetilde{Y}) \mathbin{\dot\cup} (Y^\sharp \times Y^\sharp).
$$

Note that

$$
(\widetilde{Y} \mathbin{\dot\cup} Y^\sharp) \times (\widetilde{Y} \mathbin{\dot\cup} Y^\sharp) = (\widetilde{Y} \times \widetilde{Y}) \mathbin{\dot\cup} \left[(\widetilde{Y} \times Y^\sharp) \mathbin{\dot\cup} (Y^\sharp \times \widetilde{Y}) \mathbin{\dot\cup} (Y^\sharp \times Y^\sharp) \right]
$$

since $\widetilde{Y} \times \widetilde{Y}$ and $\left[(\widetilde{Y} \times Y^\sharp) \mathbin{\dot\cup} (Y^\sharp \times \widetilde{Y}) \mathbin{\dot\cup} (Y^\sharp \times Y^\sharp) \right]$ are disjoint. We thus have a bijection

$$
\begin{array}{rccc}
\tilde{f} \mathbin{\dot\cup} h \colon & \widetilde{Y} \mathbin{\dot\cup} Y^\sharp & \longrightarrow & (\widetilde{Y} \mathbin{\dot\cup} Y^\sharp) \times (\widetilde{Y} \mathbin{\dot\cup} Y^\sharp) \\
& y' \in \widetilde{Y} & \longmapsto & \tilde{f}(y') \\
& y'' \in Y^\sharp & \longmapsto & h(y'').
\end{array}
$$

It is easy to see that $(\widetilde{Y} \,\dot{\cup}\, Y^{\sharp}, \tilde{f} \,\dot{\cup}\, h)$ is strictly greater than $(\widetilde{Y}, \tilde{f})$, a contradiction to the maximality of $(\widetilde{Y}, \tilde{f})$.

We now know that $|Y \setminus \widetilde{Y}| < |Y|$. Since $|Y| = |\widetilde{Y}| + |Y \setminus \widetilde{Y}|$, we have that $|Y| = |\widetilde{Y}|$ by (a). Hence $|Y||Y| = |\widetilde{Y}||\widetilde{Y}| = |\widetilde{Y}| = |Y|$.

Step 2. In general, we have that

$$
\begin{aligned}
|Y| &= 1|Y|, && \text{by Exercise 5,} \\
&\le |X||Y| \le |Y||Y|, && \text{by Proposition 1.5.8(d),} \\
&= |Y|, && \text{by (a).}
\end{aligned}
$$

We conclude that $|X||Y| = |Y|$. $\qquad\square$

I heartily agree with what Paul Halmos said in his book *Naïve Set Theory*, 1971, that "general set theory is pretty trivial stuff really". The only difficulty of studying set theory for the first time is to get used to the generality and abstraction. But just as Halmos said in his conclusion in the preface of the said book, "read it, absorb it, and forget it", I can't agree more.

Dimension for the infinite dimensional case

We are now ready to tackle our original problem.

Theorem 1.5.11. *Let V be a vector space over the field F. Any two bases of V are of the same cardinality.*

Proof. From Theorem 1.4.1 it remains to prove the theorem for infinite dimensional vector spaces and in this case any of the bases of V is infinite. Let B and B' both be bases of V over F. For each $v \in B$, find $w_1, w_2, \ldots, w_n$ in B' such that $v = a_1 w_1 + \cdots + a_n w_n$ where $a_i \in F \setminus \{0\}$ for all i. Remember that this expression is unique. Define

$$
S_v = \{w_1, w_2, \ldots, w_n\}.
$$

The elements of B can be generated by $\bigcup_{v \in B} S_v$, which is a subset of B'. This implies that $B' = \bigcup_{v \in B} S_v$ since B' is a minimal generating set. Hence

$$
|B'| \le \sum_{v \in B} |S_v|, \qquad \text{by Exercise 9,}
$$

$$\leq \sum_{v \in B} \omega, \qquad \text{by Exercise 8,}$$

$$= |B|\omega, \qquad \text{by Exercise 10,}$$

$$= |B|, \qquad \text{by Theorem 1.5.10(b).}$$

By symmetry we have $|B| \leq |B'|$. Thus $|B| = |B'|$. $\qquad\qquad\square$

When V is an infinite dimensional vector space over the field F, we may simply say its dimension is ∞. With the help of Theorem 1.5.11, we may also define the **dimension** of a vector space over F to be the cardinality of any of its F-bases. The dimension of V as an F-vector space is denoted by $\dim_F V$ or simply $\dim V$ when F is understood.

Example 1.5.12. Let F be a field and let x be an indeterminate over F. Then $\dim_F \bigoplus_{i=0}^{\infty} F = \omega = \dim_F F[x]$.

Exercises 1.5

For the following exercises one may assume the axiom of choice.

1. Show that the relation $\sim$ in Definition 1.5.1 is an equivalence relation.

2. Let X, Y, Z and W be sets. Suppose $X \sim Z$ and $Y \sim W$. Prove the following assertions:

   (a) $X \stackrel{\cdot}{\cup} Y \sim Z \stackrel{\cdot}{\cup} W$;

   (b) $X \times Y \sim Z \times W$.

3. Fill in the details in the proof for Proposition 1.5.8.

4. Fill in the details in the proof for Lemma 1.5.9.

5. Let X be any set. Show that $1|X| = |X|$.

6. Let X and Y be two sets. Suppose there is a surjection from X onto Y. Show that $Y \preceq X$ assuming the axiom of choice.

7. Show that both $\mathbb{Z}$ and $\mathbb{Q}$ are countable and $\mathbb{R}$ is uncountable.

8. Let $\{X_i\}_{i \in I}$ be a family of sets. Define

$$\dot{\bigcup_{i \in I}} X_i = \bigcup_{i \in I} (X_i \times \{i\}) \qquad \text{and} \qquad \sum_{i \in I} |X_i| = \left| \dot{\bigcup_{i \in I}} X_i \right|.$$

Suppose $\{Y_i\}_{i \in I}$ is another family of sets such that $|X_i| \leq |Y_i|$ for all $i \in I$. Show that

$$\sum_{i \in I} |X_i| \leq \sum_{i \in I} |Y_i|.$$

9. Suppose $\{X_i\}_{i \in I}$ is a family of subsets in the set X. Show that

$$\left| \bigcup_{i \in I} X_i \right| \leq \sum_{i \in I} |X_i|.$$

10. Let X and Y be two sets. Show that $\sum_{x \in X} |Y| = |X||Y|$.

11. Let x be a variable over the field F. Show that $|F[x]| = |F|$.

12. Let V be a field and V be an F-vector space with a finite or countable basis. Show that $|V| = |F|$ if F is infinite.

13. Show that the field of algebraic numbers ($\overline{\mathbb{Q}}$) is countable. Find $\dim_{\mathbb{Q}} \overline{\mathbb{Q}}$.

14. Let F be a field and let V and W be F-vector spaces. Show that

$$\dim(V \oplus W) = \dim V + \dim W.$$

1.6   More on cardinal arithmetic

We do not necessarily need the materials in this section for the purpose of this book. This section is really more for fun than anything else.

Basically speaking, we will discuss $|X|^{|Y|}$ in this section. For this purpose we review the formal definition of "functions".

Definition 1.6.1. Let X and Y be two sets. A **function** (or a **map** or a **mapping**) $f: X \to Y$ is a subset of $X \times Y$ such that the following conditions hold.

- For every $x \in X$, there is a $y \in Y$ such that $(x, y) \in f$.

- If $(x, y) \in f$ and $(x, z) \in f$, then $y = z$.

We call f a function from X into Y. The set X is called the **domain** of f and Y is called the **codomain** of f. If $(x, y) \in f$ we say y is the **value** of x under f and we may also write $f(x) = y$. We use $f(X)$ to denote the set

$$\{y \in Y : \text{there exists } x \in X \text{ such that } (x, y) \in f\}.$$

This is called the **range** or the **image** of f. We use Y^X to denote the set of all functions from X into Y.

The concept of "functions" thus defined is just a formal way to describe the "functions" we have been using so far in this book. As a practice, we leave it to the reader to define injective, surjective and bijective functions using the formal language in Definition 1.6.1.

Example 1.6.2. Let $X = \{a, b\}$ and $Y = \{0, 1, 2\}$. The sets

$$f_1 = \{(a, 0),\ (b, 0)\} \quad \text{and} \quad f_2 = \{(b, 1),\ (a, 2)\}$$

are two functions from X to Y. None of the following sets

$$\varnothing,\ g_1 = \{(a, 0),\ (a, 1)\},\ g_2 = \{(a, 2)\},\ \text{and}\ g_3 = \{(b, 0),\ (a, 2),\ (b, 2)\}$$

is a function from X to Y.

Example 1.6.3. Let $X = \{0, 1\}$ and $Y = \{0, 1, 2\}$. Then

$$Y^X = \{\{(0, 0), (1, 0)\},\ \{(0, 0), (1, 1)\},\ \{(0, 0), (1, 2)\}$$
$$\{(0, 1), (1, 0)\},\ \{(0, 1), (1, 1)\},\ \{(0, 1), (1, 2)\}$$
$$\{(0, 2), (1, 0)\},\ \{(0, 2), (1, 1)\},\ \{(0, 2), (1, 2)\}\}.$$

Definition 1.6.4. For the two sets X and Y, define $|X|^{|Y|} = |X^Y|$. (See Exercise 1.)

When X and Y are nonempty finite sets, we leave it to the reader to verify that $|Y|^{|X|} = |Y^X|$ in the sense taught in elementary schools.

Example 1.6.5. Let X be any nonempty set and let $x \in X$. There is no y to be found in $\varnothing$ so that $(x, y) \in f$. Hence $\varnothing^X = \varnothing$. In other words, $0^{|X|} = 0$ when $|X| \neq 0$.

On the other hand, let Y be any set. The empty set is the only subset of $\varnothing \times Y$, and $\varnothing$ automatically satisfies the requirements of a function from $\varnothing$ to Y. Hence $Y^\varnothing = \{\varnothing\}$. This argument works even when $Y = \varnothing$. Hence $|Y|^0 = 1$ for any set Y. In particular, we have $0^0 = 1$.[6]

Proposition 1.6.6. *Let X, Y, Z and W be nonempty sets. If $|X| \leq |Z|$ and $|Y| \leq |W|$, then $|Y|^{|X|} \leq |W|^{|Z|}$.*

Proof. Suppose given injections $\varphi \colon X \to Z$ and $\psi \colon Y \to W$. Let $w_0 \in W$.

We first construct a mapping Θ from Y^X to W^Z. Let $f \in Y^X$. The map Θ sends f to the set

$$\widetilde{f} = \{(z, w_0) \in Z \times W : z \notin \varphi(X)\}$$
$$\cup \{(z, \psi(y)) \in Z \times W : z = \varphi(x) \text{ for some } x \in X \text{ and } (x, y) \in f\}.$$

We claim that $\widetilde{f}$ is a function from Z to W. See Figure 1.1.

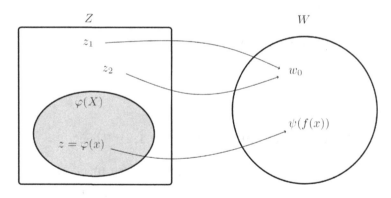

Figure 1.1: How to insert Y^X into W^Z

From the construction, for any $z \in Z$, there is a $w \in W$ such that $(z, w) \in \widetilde{f}$. The element w is either an element in $\psi(Y)$ or w_0 depending

---

[6]This statement works only in the context of set theory. When you see 0^0 elsewhere, it may mean something different.

on whether z is in $\varphi(X)$ or not. Let (z, w) and (z, w') be in $\widetilde{f}$. If $z \notin \varphi(X)$, then $w = w_0 = w'$. In case $z = \varphi(x) \in \varphi(X)$, then $w = \psi(y)$ for some $(x, y) \in f$ and $w' = \psi(y')$ for some $(x, y') \in f$. However, f is a function from X to Y. This implies that $y = y'$, and in turn $w = w'$. We have now verified that $\widetilde{f}$ is a function from Z to W. Hence Θ gives a mapping from Y^X to W^Z.

Next we show that Θ is an injection. Let f and g be two distinct functions in Y^X. Then either $f \not\subseteq g$ or $g \not\subseteq f$. Without loss of generality we may assume it is the former case. Thus there exists $(x, y) \in f \setminus g$. Find y' in Y such that $(x, y') \in g$. Clearly $y \neq y'$. We then have $(\varphi(x), \psi(y)) \in \Theta(f)$ and $(\varphi(x), \psi(y')) \in \Theta(g)$. Since ψ is an injection, we have $\psi(y) \neq \psi(y')$. If $(\varphi(x), \psi(y)) \in \Theta(g)$, we would have a contradiction to the fact that $\Theta(g)$ is a function. Thus $(\varphi(x), \psi(y)) \in \Theta(f) \setminus \Theta(g)$. We have shown that $\Theta(f) \neq \Theta(g)$. Hence Θ is indeed an injection. We conclude that $Y^X \preceq W^Z$ and $|Y|^{|X|} \leq |W|^{|Z|}$. $\qquad\square$

Example 1.6.7. It is well-known that $\omega < 2^\omega$. We will give an argument here.

For $k \in \omega$, define

$$f_k = \{(x, 0) \in \omega \times 2 : x \neq k\} \cup \{(k, 1)\}.$$

This is a function from ω to 2. It is also clear that

$$\{(k, f_k) \in \omega \times 2^\omega : k \in \omega\}$$

is a function from ω to 2^ω and is an injection. Hence $\omega \leq 2^\omega$.

We now need to show that $\omega \neq 2^\omega$, or equivalently, there is no bijection from ω to 2^ω. Suppose the opposite. Let $\Theta \colon \omega \to 2^\omega$ be a bijection. Define the function g from ω to 2 by letting

$$g(k) = \begin{cases} 0, & \text{if } \Theta(k)(k) = 1, \\ 1, & \text{if } \Theta(k)(k) = 0 \end{cases}$$

for each $k \in \omega$. Then $g \in 2^\omega$. However, $g \neq \Theta(k)$ for any k in ω since $g(k) \neq \Theta(k)(k)$. We have reached a contradiction.

Exercise 4 says that $\omega < |\mathbb{R}|$. Thus $\mathbb{R}$ is an uncountable set.

Exercises 1.6

For the following exercises one may assume the axiom of choice.

1. Let X, Y, Z and W be sets. Suppose $X \sim Z$ and $Y \sim W$. Show that $Y^X \sim W^Z$.

2. Let $X = \{0, 1, 2\}$ and $Y = \{0, 1\}$. Give all the elements in Y^X.

3. Let X be any set. Show that $1^{|X|} = 1$ and $|X|^1 = |X|$.

4. Show that $|\mathbb{R}| = 2^\omega$.

5. Let X be an infinite set. Show that $|X| < 2^{|X|}$.

6. Let $\{X_i\}_{i \in I}$ be a family of sets. Define

$$\prod_{i \in I} |X_i| = \left| \prod_{i \in I} X_i \right|.$$

Let X and Y be two sets. Show that

$$\prod_{x \in X} |Y| = |Y|^{|X|}.$$

7. Let X, Y and Z be sets. Show that $|X|^{|Y|+|Z|} = |X|^{|Y|}|X|^{|Z|}$, $(|X||Y|)^{|Z|} = |X|^{|Z|}|Y|^{|Z|}$ and $(|X|^{|Y|})^{|Z|} = |X|^{|Y||Z|}$.

8. Suppose that X is a finite nonempty set and Y is an infinite set. Show that $|X|^{|Y|} = \omega^{|Y|}$. (Hint: $|X|^{|Y|} = |X|^{\omega|Y|}$.)

CHAPTER 2

Linear Maps

A module (or a vector space) is equipped with a "vector" addition and a "scalar" multiplication. A *morphism* between modules must preserve the two operations. Such a morphism will be called a *linear map*. The more familiar term *linear transformation* will be replaced by the general term *linear map* in this book as well.

Besides discussing the properties of a linear map, we will also discuss *quotient modules* and the *fundamental theorem of linear maps*. Using the fundamental theorem enables us to determine the structure (or isomorphic classes) of modules. In particular, in the case of vector spaces, *dimension* is the key factor that determines whether two vector spaces are isomorphic to each other. The *rank* and *nullity* of a linear transformation are examples that demonstrate the power of bases and dimension.

A linear map on free modules of finite rank may be associated with a matrix with entries in the ring (of field) of scalars. Sometimes it is best to approach a linear map through its matrix. And sometimes it is easier to understand a matrix by studying the linear map it represents. The study of linear maps and the study of matrices are two sides of a coin.

2.1 Linear maps

Throughout this section R denotes a ring and F denotes a field.

Linear maps and isomorphisms

Definition 2.1.1. Let M and N be R-modules. A function $f\colon M \to N$ is called an R-**linear map** if

$$f(m + m') = f(m) + f(m') \qquad \text{and} \qquad f(rm) = r f(m)$$

for all $r \in R$ and m, $m' \in M$.

If V and W are F-vector spaces and $T\colon V \to W$ is an F-linear map, we usually say that T is an F-**linear transformation**.

Linear maps are the homomorphisms among modules. The following result is an alternative method to verify whether a function is a linear map.

Lemma 2.1.2. *Let M and N be R-modules. The function $f\colon M \to N$ is an R-linear map if and only if*

$$f(rm + m') = r f(m) + f(m')$$

for all $r \in R$ and m, $m' \in M$.

We leave the proof as an easy exercise. See Exercise 1.

Example 2.1.3. Let M and N be two R-modules. The **zero map** or the **trivial map** sending every element in M to the additive identity 0_N in N is an R-linear map. The **identity map** 1_M is also an R-linear map from M to itself.

Example 2.1.4. Let x be a variable over $\mathbb{R}$. Consider the differential map D from $\mathbb{R}[x]$ to itself sending $f(x)$ to $f'(x)$. This is an $\mathbb{R}$-linear transformation. So is the integration map sending $f(x)$ to $\int_0^x f(t)\, dt$.

Definition 2.1.5. Let $f\colon M \to N$ be an R-linear map. Just as what we had done before, we will use $\mathrm{Im}\, f$ or $f(M)$ to denote the **image** of f. We call the set $f^{-1}(0)$ the **kernel** of f, denoted $\mathrm{Ker}\, f$.

It is routine to verify the following basic properties, and we leave them as exercises. See Exercises 2 and 3.

Lemma 2.1.6. *Let $f\colon M \to N$ be an R-linear map between R-modules. Then $f(0_M) = 0_N$. The image $\operatorname{Im} f$ is a submodule of N and the kernel $\operatorname{Ker} f$ is a submodule of M.*

Lemma 2.1.7. *Let L, M and N be R-modules and let $f\colon L \to M$ and $g\colon M \to N$ be R-linear maps. Then $gf\colon L \to N$ is also an R-linear map.*

Definition 2.1.8. A bijective linear map is called an **isomorphism**. We say the R-module M is **isomorphic to** the R-module N, denoted $M \cong N$, if there is an isomorphism from M to N.

Lemma 2.1.9. *Let $f\colon M \to N$ be an isomorphism between the R-modules M and N.*

(a) *The inverse function of f is also an isomorphism.*

(b) *"Isomorphic to" is an equivalence relation.*

Proof. (a) Remember that a function has an inverse if and only if it is bijective. This says that f^{-1} is bijective since f is the inverse of f^{-1}. Thus it suffices to show that the inverse $f^{-1}\colon N \to M$ is also an R-linear map.

Suppose given n, $n' \in N$ and $r \in R$. Find m and $m' \in M$ such that $f(m) = n$ and $f(m') = n'$. Since f is an R-linear map, we have that $f(m + m') = f(m) + f(m') = n + n'$ and $f(rm) = rf(m) = rn$. Thus

$$f^{-1}(n + n') = m + m' = f^{-1}(n) + f^{-1}(n');$$
$$f^{-1}(rn) = rm = rf^{-1}(n).$$

The inverse f^{-1} is R-linear.

(b) Let L, M and N be R-modules. The identity map 1_M is an isomorphism. It implies that $M \cong M$. If $M \cong N$, let $f\colon M \to N$ be an isomorphism. The inverse $f^{-1}\colon N \to M$ is also an isomorphism from (a). We have $N \cong M$. Finally, suppose $L \cong M$ and $M \cong N$. Let $f\colon L \to M$ and $g\colon M \to N$ be isomorphisms. Then gf is bijective and R-linear by Lemma 2.1.7. Thus fg is an isomorphism and $L \cong N$. We conclude that $\cong$ is an equivalence relation. $\qquad\square$

When two modules or two vector spaces are isomorphic, basically the two modules (or vector spaces respectively) are the same in terms of properties regarding addition and scalar multiplication. The following result is just one example.

Proposition 2.1.10. *Let R be a ring and M be a free R-module. Suppose given a basis B for M over R. Let $f\colon M \to N$ be an isomorphism. Then $B' = f(B) = \{f(u) \in N : u \in B\}$ is also a basis for N over R. Thus N is a free R-module as well.*

Proof. Let $n \in N$. Find $m \in M$ such that $f(m) = n$. Find $u_1, u_2, \ldots, u_s$ in B so that $m = \sum_{i=1}^{s} a_i u_i$ where $a_i \in R$. Hence $n = f(m) = \sum_{i=1}^{s} a_i f(u_i)$. This implies that B' generates N.

Let $f(u_1), f(u_2), \ldots, f(u_s) \in B'$ where $u_1, u_2, \ldots, u_s$ are distinct elements in B. Let $a_1, a_2, \ldots, a_s \in R$ be such that $\sum_{i=1}^{s} a_i f(u_i) = 0$. Then $f\left(\sum_{i=1}^{s} a_i u_i\right) = \sum_{i=1}^{s} a_i f(u_i) = 0$. Since f is injective, we have that $\sum_{i=1}^{s} a_i u_i = 0$ in M. This implies that $a_1, a_2, \ldots, a_s$ are all 0 since B is linearly independent over R. This shows that B' is also linearly independent over R. Hence N contains B' as a basis and N is a free R-module. $\square$

Free modules

The most important property regarding free modules are given in the following theorem. This theorem says that a free module may be mapped into an arbitrary R-module by "freely" assigning values to a given basis.

Theorem 2.1.11. *Let $B = \{u_i\}_{i \in I}$ be a basis for the free module M over R. For each $i \in I$, assign an element n_i in the R-module N. There is a unique R-linear map from M to N sending u_i to n_i.*

Proof. Define a function $f\colon M \to N$ in the following manner. For each $m \in M$, there is a unique expression for m as an R-linear combination of elements in B, say $m = a_1 u_{i_1} + a_2 u_{i_2} + \cdots + a_s u_{i_s}$ where $a_1, a_2, \ldots, a_s \in R$ and $u_{i_1}, u_{i_2}, \ldots, u_{i_s}$ are distinct elements in B. Define

$$f(m) = a_1 n_{i_1} + a_2 n_{i_2} + \cdots + a_s n_{i_s}.$$

We need to verify that f is indeed R-linear.

Let m, $m' \in M$ and let $r \in R$. It is possible to find a finite subset $\{u_{i_1}, u_{i_2}, \ldots, u_{i_s}\}$ within B such that

$$m = a_1 u_{i_1} + a_2 u_{i_2} + \cdots + a_s u_{i_s}$$
$$m' = a'_1 u_{i_1} + a'_2 u_{i_2} + \cdots + a'_s u_{i_s}$$

where $a_1, a_2, \ldots, a_s, a'_1, a'_2, \ldots, a'_s \in R$. Then

$$
\begin{aligned}
f(rm + m') &= f((ra_1 + a'_1)u_{i_1} + (ra_2 + a'_2)u_{i_2} + \cdots + (ra_s + a'_s)u_{i_s}) \\
&= (ra_1 + a'_1)n_{i_1} + (ra_2 + a'_2)n_{i_2} + \cdots + (ra_s + a'_s)n_{i_s} \\
&= \sum_{j=1}^{s} ra_j n_{i_j} + \sum_{j=1}^{s} a'_j n_{i_j} = r \sum_{j=1}^{s} a_j n_{i_j} + \sum_{j=1}^{s} a'_j n_{i_j} \\
&= rf(m) + f(m').
\end{aligned}
$$

Thus f is R-linear by Lemma 2.1.2.

Suppose that $g \colon M \to N$ is any R-linear map sending u_i to n_i for each $i \in I$. Let $m = a_1 u_{i_1} + a_2 u_{i_2} + \cdots + a_s u_{i_s} \in M$. Then

$$g(m) = \prod_{j=1}^{s} a_i g(u_{i_j}) = \prod_{j=1}^{s} a_i n_{i_j} = f(m).$$

This is true for all $m \in M$. The uniqueness of f is thus established. $\qquad\square$

Corollary 2.1.12. *Let M and N be free R-modules with bases B and C respectively. If $|B| = |C|$, then $M \cong N$.*

Proof. Let φ be a bijection from B to C. From Theorem 2.1.11, φ can be extended to an R-linear map $f \colon M \to N$. Similarly, there is an R-linear map $g \colon N \to M$ extending φ^{-1}. Note that gf maps any element in B to itself. The identity map $\mathbf{1}_M$ also maps any element in B to itself. By the uniqueness part of Theorem 2.1.11 we have that $gf = \mathbf{1}_M$. Similarly, $fg = \mathbf{1}_N$. Hence f is an isomorphism and $M \cong N$. $\qquad\square$

Corollary 2.1.13. *Two F-vector spaces are isomorphic if and only if they are of the same dimension (in terms of cardinality of a basis).*

In particular, $V \cong F^n$ if V is an n-dimensional vector space over F.

Proof. Let V and W be two vector spaces such that $\dim V = \dim W$.

The "if" part follows from Corollary 2.1.12.

The "only if" part: Let $f \colon V \to W$ be an isomorphism. Let B be a basis for V over F. We have that $f(B)$ is a basis for W over F by Proposition 2.1.10. The isomorphism induces a bijection from B to $f(B)$. Hence $\dim V = \dim W$.

The last statement follows from the fact $\dim_F F^n = n$. See Example 1.3.5.   □

Example 2.1.14. Let $B = \{u_i\}_{i=1}^n$ be a basis for M over R. The isomorphism from M to R^n sending u_i to e_i gives a correspondence between $m = a_1 u_1 + \cdots + a_n u_n$ in M and $a_1 e_1 + \cdots + a_n e_n = (a_1, a_2, \ldots, a_n)$ in R^n.

Example 2.1.15. Recall $\mathscr{P}_n \subseteq F[x]$ where x is an indeterminate over F. It is an $(n+1)$-dimensional vector space over F with $\{1, x, x^2, \ldots, x^n\}$ as the standard basis. The polynomial $a_0 + a_1 x + \cdots + a_n x^n$ in $F[x]$ may correspond to the vector $(a_0, a_1, a_2, \ldots, a_n)$ in F^{n+1}.

Example 2.1.16. Let $S = \{1, 2, 3\}$. Consider the $\mathbb{R}$-vector space $\mathbb{R}^S$ and its standard basis $\{\chi_1, \chi_2, \chi_3\}$. For example, the function

$$f : 1 \mapsto 2, \quad 2 \mapsto -3, \quad 3 \mapsto \sqrt{2}$$

can be expressed as $2\chi_1 - 3\chi_2 + \sqrt{2}\chi_3$. Thus f may correspond to $(2, -3, \sqrt{2})$ in $\mathbb{R}^3$.

Exercises 2.1

Throughout these exercises, R denotes a ring and F denotes a field.

1. Prove Lemma 2.1.2.

2. Prove Lemma 2.1.6.

3. Prove Lemma 2.1.7.

4. Let L, M and N be R-modules. Show that $M \oplus N \cong N \oplus M$ and
   $(L \oplus M) \oplus N \cong L \oplus (M \oplus N)$. See Exercise 7, §1.1.

5. Let M and N be free modules over R. Let B be a basis for M
   over R. Show that $f \colon M \to N$ is an isomorphism if and only if
   $\{f(b) \in N : b \in B\}$ is a basis for N over R.

6. Let I be an ideal of R and M be an R/I-module. Remember that M
   may be viewed as an R-module too (see Example 1.1.12).

   (a) If f is an R/I-linear map, show that f is also R-linear.

   (b) If f is an R-linear map, show that f is also (R/I)-linear.

7. Let M and N be R-modules. Denote by $\operatorname{Hom}_R(M, N)$ the set of all
   R-linear maps from M to N.

   (a) Let f and $g \in \operatorname{Hom}_R(M, N)$ and let $a, b \in R$. Show that the
   function $h \colon M \to N$ defined by letting

   $$h(m) = af(m) + bg(m)$$

   is an R-linear map. We usually write $h = af + bg$. Thus there
   is a natural addition and a natural scalar multiplication inside
   $\operatorname{Hom}_R(M, N)$.

   (b) Show that $\operatorname{Hom}_R(M, N)$ is an R-module via (a).

8. Let M, N, M' and N' be R-modules such that $M \cong M'$ and $N \cong N'$.
   Show that $\operatorname{Hom}_R(M, N) \cong \operatorname{Hom}_R(M', N')$.

9. Let $\{u_i : i \in I\}$ and $\{v_j : j \in J\}$ be bases for the R-modules M and
   N respectively. Let $s \in I$ and $t \in J$. Define

   $$f_{st}(u_i) = \begin{cases} v_t, & i = s, \\ 0, & i \in I \setminus \{s\}. \end{cases}$$

   (a) Show that

   $$B = \{f_{st} \in \operatorname{Hom}_R(M, N) : s \in I,\ t \in J\}$$

   is linearly independent over R.

(b) Show that B generates $\operatorname{Hom}_R(M, N)$ over R when M is free of finite rank.

(c) Does B still generate $\operatorname{Hom}_R(M, N)$ over R when only N is free of finite rank?

(d) When M and N are both free modules of finite rank, show that $\operatorname{Hom}_R(M, N)$ is free of finite rank and find its rank over R.

10. Let V and W be finite dimensional F-vector spaces. Find the dimension of $\operatorname{Hom}_F(V, W)$ over F.

11. Let L and M be R-modules and N be a submodule of L over R. Let

$$\mathscr{U} = \{f \in \operatorname{Hom}_R(L, M) : f(n) = 0 \text{ for all } n \in N\}.$$

Show that $\mathscr{U}$ is a submodule of $\operatorname{Hom}_R(L, M)$ over R.

12. Let L, M and N be R-modules.

(a) Is it true that $\operatorname{Hom}_R(L \oplus M, N) \cong \operatorname{Hom}_R(L, N) \oplus \operatorname{Hom}_R(M, N)$?

(b) Is it true that $\operatorname{Hom}_R(L, M \oplus N) \cong \operatorname{Hom}_R(L, M) \oplus \operatorname{Hom}_R(L, N)$?

13. Let U, V and W be F-vector spaces.

(a) Is it true that $\operatorname{Hom}_F(U \oplus V, W) \cong \operatorname{Hom}_F(U, W) \oplus \operatorname{Hom}_F(V, W)$?

(b) Is it true that $\operatorname{Hom}_F(U, V \oplus W) \cong \operatorname{Hom}_F(U, V) \oplus \operatorname{Hom}_F(U, W)$?

14. Let M be an R-module. Show that $\operatorname{Hom}_R(M, M)$ is a (usually noncommutative) ring if the product is given by

$$(fg)(m) = (f \circ g)(m)$$

for $f, g \in \operatorname{Hom}_R(M, M)$.

Let $f \colon M \to M$ be an R-linear map. Show that $R[f]$ is a commutative subring of $\operatorname{Hom}_R(M, M)$.

15. Let V be an F-vector space. Then $\operatorname{Hom}_F(V, F)$ is also called the **dual space** of V over F. Any element in $\operatorname{Hom}_F(V, F)$ is called a **linear functional** on V over F. More generally, if M is an R-module,

$\mathrm{Hom}_R(M, R)$ is called the **dual module** of M over R. The dual space and the dual module are typically denoted as V^* and M^* respectively.

Let M be a free module of finite rank over R. Suppose given a basis $B = \{u_1, u_2, \ldots, u_n\}$ for M over R. Show that B gives rise to a basis $B' = \{f_1, f_2, \ldots, f_n\}$ for M^* where

$$f_j(u_i) = \delta_{ij} = \begin{cases} 1, & \text{if } i = j; \\ 0, & \text{otherwise.} \end{cases}$$

The basis B' is called the **dual basis** of B.

16. Suppose given an inner product $\langle -, - \rangle$ in $\mathbb{R}^n$. Let $v_0 \in \mathbb{R}^n$. Show that $f \colon \mathbb{R}^n \to \mathbb{R}$ sending $w \in \mathbb{R}^n$ to $\langle v_0, w \rangle$ is a linear functional on $\mathbb{R}^n$ over $\mathbb{R}$.

17. Let $\langle -, - \rangle$ be the standard inner product with respect to the standard basis $\{e_1, e_2, e_3\}$ for $\mathbb{R}^3$. Let $f = \langle (1, 2, -3), - \rangle \in \mathrm{Hom}_{\mathbb{R}}(\mathbb{R}^3, \mathbb{R})$. Express f as a linear combination of elements in the dual basis of $\{e_1, e_2, e_3\}$.

18. Let V be an F-vector space with a basis $\{v_1, v_2, \ldots, v_n\}$. For any given n-tuple $(a_1, a_2, \ldots, a_n)$ in F^n, show that there is a unique linear functional f on V such that $f(v_i) = a_i$.

19. Let u and v be two distinct vectors in the finite dimensional vector space V. Show that there is a linear functional f on V such that $f(u) \neq f(v)$.

2.2 Quotient modules and quotient spaces

Throughout this section R denotes a ring and F denotes a field.

Quotient modules

Let M be an R-module and N be a submodule of M. Additively, N is a (normal) subgroup of M. Hence M/N is also an additive group. A typical element in M/N is denoted as $m + N$ or simply $\overline{m}$ where $m \in M$. Now notice that there is a natural scalar product of R on M/N given by

$$r \cdot \overline{m} = \overline{rm}.$$

This is feasible since $r(m + N) = rm + rN \subseteq rm + N$. Thus, M/N is naturally an R-module. This is called the **quotient module** of M modulo N. If W is a subspace of the F-vector space V, V/W is called the **quotient space** of V modulo W.

Example 2.2.1. Let L be the line $x = y$ in $\mathbb{R}^2$, which is a one-dimensional subspace of $\mathbb{R}^2$. Then V/W consists of lines parallel to L. In fact, the coset $v + L$ is the line parallel to L and passing through (the tip of) v. See Figure 2.1.

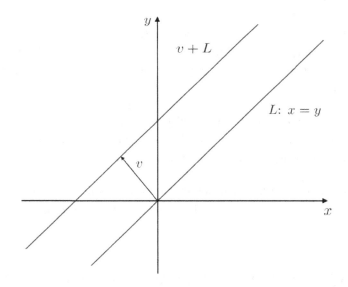

Figure 2.1: The quotient space of $\mathbb{R}^2/L$
consists of lines parallel to L.

Lemma 2.2.2. *Let M be an R-module and let N be a submodule of M. Suppose $N = \langle n_i : i \in I \rangle$ and $M/N = \langle \overline{m}_j : j \in J \rangle$ where the m_j's are in M. Then $M = \langle A \cup B \rangle$ where $A = \{ n_i : i \in I \}$ and $B = \{ m_j : j \in J \}$.*

Proof. Let $m \in M$. We may find $a_1, a_2, \ldots, a_r \in R$ such that

$$\overline{m} = \sum_{k=1}^{r} a_k \overline{m}_{j_k}$$

in M/N. This implies that $m - \sum_{k=1}^{r} a_k m_{j_k} \in N$. Hence we may find $b_1, b_2, \ldots, b_s \in R$ such that $m - \sum_{k=1}^{r} a_k m_{j_k} = \sum_{\ell=1}^{s} b_\ell n_{i_\ell}$. Thus

$$m = \sum_{k=1}^{r} a_k m_{j_k} + \sum_{\ell=1}^{s} b_\ell n_{i_\ell}$$

is generated by $A \cup B$. □

Proposition 2.2.3. *Let M be an R-module and let N be a submodule of M. Suppose that $A = \{u_i\}_{i \in I}$ is a basis of N over R and $B = \{v_j\}_{j \in J}$ is a subset of M such that $\overline{B} = \{\overline{v_j}\}_{j \in J}$ is a basis of M/N over R. Then $A \cap B = \varnothing$ and $A \cup B$ is a basis of M over R.*

Thus, if N and M/N are both free modules, so is M.

Remark. Please go to Exercise 1 to see the converse of the last statement is not true.

Proof. If $m \in A \cap B$, then $\overline{m} = 0$ in M/N while $\overline{m}$ belongs to a basis for M/N, a contradiction. Hence $A \cap B = \varnothing$.

Suppose

$$\sum_{k=1}^{r} a_k v_{j_k} + \sum_{\ell=1}^{s} b_\ell u_{i_\ell} = 0, \qquad a_k, b_\ell \in R \text{ for all } k, \ell.$$

After modulo N we obtain $\sum_{k=1}^{r} a_k \overline{v_{j_k}} = 0$. We have that the a_k's are all zero since $\overline{B}$ is a basis for M/N. It follows that $\sum_{\ell=1}^{s} b_\ell u_{i_\ell} = 0$. Then b_ℓ's are all trivial since A is a basis for N. This shows that $A \cup B$ is linearly independent over R. We now conclude that $A \cup B$ is a basis for M over R from Lemma 2.2.2, and the last statement is an immediate result. □

The following proposition is also an immediate result of Proposition 2.2.3 since everything is free in the case of vector spaces.

Corollary 2.2.4. *Let V be an F-vector space and let W be a subspace of V. Then*

$$\dim V = \dim W + \dim V/W.$$

In particular, if V is finite dimensional, we have that

$$\dim \frac{V}{W} = \dim V - \dim W.$$

Fundamental theorem of homomorphisms

Let M be an R-module and let N be a submodule of M. There is a **canonical epimorphism**

$$\begin{array}{rccc} \pi : & M & \longrightarrow & M/N \\ & m & \longmapsto & \overline{m}. \end{array}$$

This is easily seen to be R-linear:

$$\pi(rm + m') = \overline{rm + m'} = r\overline{m} + \overline{m'} = r\pi(m) + \pi(m')$$

for $r \in R$ and $m,\, m' \in M$. It is also clear that π is surjective, and hence an epimorphism from construction.

We first review the fundamental theorem of homomorphisms for groups. The reader may find the proof and discussion of this theorem in any undergraduate textbook for abstract algebra.

Theorem 2.2.5 (Fundamental theorem of group homomorphisms). *Let $\eta\colon G \to H$ be a group homomorphism between groups and let $K \subseteq \operatorname{Ker} \eta$ be a normal subgroup of G. Let $\pi\colon G \twoheadrightarrow G/K$ be the canonical epimorphism. There is a unique group homomorphism $\widetilde{\eta}\colon G/K \to H$ such that $\eta = \widetilde{\eta}\pi$.*

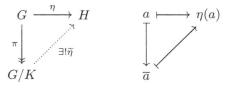

In particular, $\widetilde{\eta}$ is a monomorphism if and only if $K = \operatorname{Ker} \eta$.

We may use the group version of the fundamental theorem of homomorphisms to obtain the following result.

Theorem 2.2.6 (Fundamental theorem of R-linear maps). *Let $f\colon M \to N$ be an R-linear map of R-modules and let $K \subseteq \operatorname{Ker} f$ be a submodule of M. Let $\pi\colon M \twoheadrightarrow M/K$ be the canonical epimorphism. There exists a unique R-linear map $\widetilde{f}\colon M/K \to N$ such that $f = \widetilde{f}\pi$.*

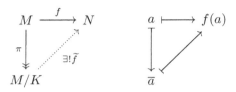

In particular, $\widetilde{f}$ is a monomorphism if and only if $K = \operatorname{Ker} f$.

Proof. The module M is by definition an additive (abelian) group. Hence the submodule K is by default a normal subgroup. Thus this theorem holds at least at the group level: there exists a unique map $\widetilde{f}\colon M/K \to N$ such that $f = \widetilde{f}\pi$ and $\widetilde{f}(\overline{m} + \overline{m'}) = \widetilde{f}(\overline{m}) + \widetilde{f}(\overline{m'})$ for m, $m' \in M$. For $r \in R$ and $m \in M$ we also have

$$\widetilde{f}(r\overline{m}) = \widetilde{f}(\overline{rm}) = \widetilde{f}(\pi(rm)) = f(rm) = rf(m) = r\widetilde{f}(\pi(m)) = r\widetilde{f}(\overline{m}),$$

since f is R-linear. Thus $\widetilde{f}$ is an R-linear map. The rest of the theorem follows from Theorem 2.2.5. $\qquad\qquad\square$

　　Just as the group version, Theorem 2.2.6 has the same set of corollaries as follows.

Theorem 2.2.7 (Correspondence theorem). *Let M be an R-module, N be a submodule of M and let $\pi\colon M \to M/N$ be the canonical epimorphism. Let $\mathscr{A}$ be the family of submodules of M containing N and let $\mathscr{B}$ be the family of submodules of M/N. There is an order-preserving one-to-one correspondence between*

$$
\begin{array}{ccc}
\mathscr{A} & \longleftrightarrow & \mathscr{B} \\
K & \longmapsto & \overline{K} = K/N \\
\pi^{-1}(\mathbb{K}) & \longleftarrow\!\!\!\mapsto & \mathbb{K}
\end{array}
$$

The following are the *three isomorphism theorems.*

Theorem 2.2.8 (The first isomorphism theorem for modules). *Suppose f is an epimorphism from M onto N. Then*

$$\frac{M}{\operatorname{Ker} f} \cong N.$$

Theorem 2.2.9 (The second isomorphism theorem for modules). *Let N and K be submodules of the R-module M. Then*

$$\frac{N + K}{K} \cong \frac{N}{N \cap K}.$$

Theorem 2.2.10 (The third isomorphism theorem for modules). *Suppose $K \subseteq N$ are submodules of the R-module M. Then N/K is also a submodule of M/K and*

$$\frac{M/K}{N/K} \cong \frac{M}{N}.$$

We may also use the group version of Theorems 2.2.7–2.2.10 to help prove these theorems like we did in the proof of Theorem 2.2.6. We leave the proof of these theorems to the reader. See Exercises 2–5.

We may use the isomorphism theorems to prove the following well-known result regarding the dimension of the intersection of two subspaces.

Example 2.2.11. Let's revisit Exercise 7, §1.4. Consider the linear transformation $T\colon V \oplus W \to V + W$ sending (v, w) to $v + w$. It is clear that T is F-linear and surjective. By the first isomorphism theorem, we have

$$V + W \cong \frac{V \oplus W}{\operatorname{Ker} T}.$$

We claim that

$$\operatorname{Ker} T = \{(v, -v) : v \in V \cap W\}.$$

The "$\supseteq$" part is clear. Now let $(v, w) \in \operatorname{Ker} T$ where $v \in V$ and $w \in W$. Then

$$v + w = 0 \implies w = -v \in V \cap W.$$

Hence $(v, w) = (v, -v)$ where $v \in V \cap W$. This shows that the "$\subseteq$" part also holds. It is trivial to show that

$$
\begin{aligned}
V \cap W &\longrightarrow \operatorname{Ker} T \\
v &\longmapsto (v, -v)
\end{aligned}
$$

is an isomorphism. Hence

$$\begin{aligned} \dim(V + W) &= \dim \frac{V \oplus W}{\operatorname{Ker} T} = \dim(V \oplus W) - \dim \operatorname{Ker} T \\ &= \dim V + \dim W - \dim(V \cap W) \end{aligned}$$

by Exercise 8, §1.4 and Corollary 2.2.4.

Rank and nullity

Let $T\colon V \to W$ be a linear transformation between two F-vector spaces. The dimension of $\operatorname{Im} T$ is also called the **rank** of T and denoted $\rho(T)$. Using the terminology of linear algebra, $\operatorname{Ker} T$ is also called the **nullspace** of T. The dimension of $\operatorname{Ker} T$ is called the **nullity** of T and denoted $\mu(T)$.

We describe how to find the rank and nullity of a linear transformation. Suppose given an F-linear transformation $T\colon V \to W$. Let $\{v_1, v_2, \ldots, v_n\}$ be a basis of V over F. Then

$$\operatorname{Im} T = \operatorname{Sp}\{T(v_1), \ldots, T(v_n)\}.$$

(See Exercise 6.) We can reduce this generating set to a basis for $\operatorname{Im} T$ to obtain the rank of T. We can then use the following formula to find the nullity of T.

Proposition 2.2.12. *Let $T\colon V \to W$ be a linear transformation between two F-vector spaces. Then*

$$\rho(T) + \mu(T) = \dim V.$$

Proof. Using the first isomorphism theorem we have that $\operatorname{Im} T \cong V/\operatorname{Ker} T$. Hence

$$\begin{aligned} \dim V &= \dim \operatorname{Ker} T + \dim V/\operatorname{Ker} T \\ &= \mu(T) + \dim \operatorname{Im} T = \mu(T) + \rho(T) \end{aligned}$$

by Corollary 2.2.4. $\qquad\square$

Suppose given a homogeneous system of linear equations

$$\begin{cases} a_{11}x_1 + a_{12}x_2 + \cdots + a_{1n}x_n = 0, \\ a_{21}x_1 + a_{22}x_2 + \cdots + a_{2n}x_n = 0, \\ \quad\quad\quad\quad \vdots \quad\quad\quad\quad \vdots \\ a_{m1}x_1 + a_{m2}x_2 + \cdots + a_{mn}x_n = 0 \end{cases}$$

over the ring R. To solve this system, we may consider the R-linear map

(2.2.2)
$$\begin{array}{ccc} f: & R^n & \longrightarrow & R^m \\ & (\alpha_1, \alpha_2, \ldots, \alpha_n) & \longmapsto & (\sum_j a_{1j}\alpha_j, \ldots, \sum_j a_{mj}\alpha_j). \end{array}$$

The solution set of (2.2.1) is $\operatorname{Ker} T$, a submodule of R^n. When $R = F$ is a field, the situation is straightforward and studied in an undergraduate course of linear algebra.

Example 2.2.13. Find the dimension of the solution space of the system of linear equations
$$\begin{cases} 3x + 4y - 2z = 0, \\ 2x - 2y + 3z = 0 \end{cases}$$
over $\mathbb{Q}$.

Solution. The dimension of the solution space is the nullity of the linear transformation
$$\begin{array}{ccc} T: & \mathbb{Q}^3 & \longrightarrow & \mathbb{Q}^2 \\ & (x, y, z) & \longmapsto & (3x + 4y - 2z, 2x - 2y + 3z). \end{array}$$

We may use the standard basis $\{e_1, e_2, e_3\}$ to find $\rho(T)$. We have that

$$\operatorname{Im} T = \operatorname{Sp}\{T(e_1),\ T(e_2),\ T(e_3)\} = \operatorname{Sp}\{(3,2),\ (4,-2),\ (-2,3)\} = \mathbb{Q}^2.$$

Thus $\mu(T) = 3 - \rho(T) = 3 - 2 = 1.$ ◇

Let's return to the problem of solving a general system of linear equations

(2.2.3)
$$\begin{cases} a_{11}x_1 + a_{12}x_2 + \cdots + a_{1n}x_n = b_1, \\ a_{21}x_1 + a_{22}x_2 + \cdots + a_{2n}x_n = b_2, \\ \quad\quad\quad\quad \vdots \quad\quad\quad\quad \vdots \\ a_{m1}x_1 + a_{m2}x_2 + \cdots + a_{mn}x_n = b_m \end{cases}$$

over R. We first find a special solution $c = (c_1, c_2, \ldots, c_n)$ for (2.2.3) if it is consistent. Let T be the linear transformation given in (2.2.2). We claim that the solution set is $c + \operatorname{Ker} T$. Let $v \in \operatorname{Ker} T$. Then

$$T(c + v) = T(c) + T(v) = T(c) = b = (b_1, b_2, \ldots, b_m).$$

Hence $c + v$ is a solution of (2.2.3). Conversely, if c' is also a solution of (2.2.3), we have that $T(c' - c) = T(c') - T(c) = b - b = 0$. Thus $c' - c \in \operatorname{Ker} T$ and $c' = c + (c' - c) \in c + \operatorname{Ker} T$. Our claim is justified.

Exercises 2.2

Throughout this set of exercises R denotes a ring and F denotes a field.

1. For each of the following cases give an example of a triple (R, M, N) where M is a free module over R and N is a submodule of M.

    (a) The submodule N is not free over R.

    (b) The submodule N is free over R but the quotient module M/N is not.

2. Let M be an R-module and let $N \subseteq K$ be submodules of M. Show that K/N is a submodule of M/N. Use this result to prove Theorem 2.2.7.

3. Prove Theorem 2.2.8.

4. Prove Theorem 2.2.9.

5. Prove Theorem 2.2.10.

6. Let M and N be R-modules and let $f \colon M \to N$ be an R-linear map. Show that $\operatorname{Im} f = \langle f(m_1), f(m_2), \ldots, f(m_s) \rangle$ if M is generated by $m_1, m_2, \ldots, m_s$ over R.

7. Prove Proposition 2.2.12 without using the isomorphism theorems.

8. Let M and N be submodules of L over R.

(a) Show that the following conditions are equivalent:

    (i) $L = M + N$ and $M \cap N = \{0\}$;

    (ii) $L = M + N$ and M, N are independent over R;

    (iii) every element in L can be written in the form $l = m + n$, $m \in M$ and $n \in N$, in one and only one way.

(b) Let M and N satisfy one of the three equivalent conditions in part (a). Show that the function

$$
\begin{array}{rccc}
f: & M \oplus N & \longrightarrow & L \\
   & (m, n) & \longmapsto & m + n
\end{array}
$$

is an isomorphism.

Thus, if we have a triple (L, M, N) satisfying one of the equivalent conditions in (a), we say L is the **internal direct sum** of M and N, and we denote it by writing $L = M \oplus N$. Compare with the direct sum defined in Exercise 7, §1.1. The old direct sum is called the **external direct sum**. However, usually we will use the term "direct sum" without specifying whether the direct sum is internal or external, for which the reason is given in the next exercise.

9. Let M_1, M_2, ..., M_n be submodules of an R-module L. Suppose the following conditions hold:

(a) $L = M_1 + M_2 + \cdots + M_n$;

(b) the M_i's satisfies the *triangular* condition

$$M_1 \cap M_2 = \{0\}$$
$$(M_1 + M_2) \cap M_3 = (0)$$
$$\vdots$$
$$(M_1 + M_2 + \cdots + M_{n-1}) \cap M_n = (0).$$

Show that L is the internal direct sum $M_1 \oplus M_2 \oplus \cdots \oplus M_n$ in the sense of Exercise 6, §1.2. Show that L is also isomorphic to the external direct sum $M_1 \oplus M_2 \oplus \cdots \oplus M_n$ in the sense of Exercise 7, §1.1.

10. Let $\{v_1, v_2, \ldots, v_n\}$ be a basis of F^n over F. Show that

$$V \cong Fv_1 \oplus \cdots \oplus Fv_n \cong F^n.$$

11. In which of the following cases is it true that $\mathbb{C}^4 = \mathrm{Sp}\{x, y\} \oplus \mathrm{Sp}\{u, v\}$?

    (a) $x = (1, 1, 0, 0)$ $y = (1, 0, 1, 0)$; $u = (0, 1, 0, 1)$, $v = (0, 0, 1, 1)$.

    (b) $x = (-1, 1, 1, 0)$, $y = (0, -1, 1, 1)$; $u = (1, 0, 0, 0)$, $v = (0, 0, 0, 1)$.

    (c) $x = (1, 0, 0, 1)$, $y = (0, 1, 1, 0)$; $u = (1, 0, 1, 0)$, $v = (0, 1, 0, 1)$.

12. Let x be an indeterminate over F. Find $\dim_F F[x]/W$ for the following choices of W.

    (a) The subspace $W = \mathscr{P}_n$, the subspace of all polynomials of degree $\leq n$.

    (b) The subspace $W = \mathscr{E}_n$, the subspace of polynomials with only even degree terms.

    (c) The subspace $W = F[x]x^n$, the subspace of all polynomials divisible by x^n.

13. Let U, V, W and X be finite dimensional vector spaces over F. Let $T: V \to W$, $T_1: U \to V$ and $T_2: W \to X$ be linear transformations over F.

    (a) Show that $\rho(TT_1) \leq \rho(T)$ and $\rho(T_2T) \leq \rho(T)$.

    (b) Show that $\rho(TT_1) = \rho(T)$ if T_1 is an isomorphism.

    (c) Show that $\rho(T_2T) = \rho(T)$ if T_2 is an isomorphism.

14. Let I be an arbitrary (and possibly infinite) index set. Let N be an R-module and let $\{M_i\}_{i \in I}$ be a family of R-modules.

    (a) Show that $\bigoplus_{i \in I} M_i$ is generated by the set of elements with only one nonzero coordinate over R.

    (b) Suppose $f_i: M_i \to N$ is an R-linear map for each $i \in I$. Let $(m_i)_i \in \bigoplus_{i \in I} M_i$. The notation $\sum_{i \in I} f_i(m_i)$ makes sense since $f_i(m_i) = 0$ except for finitely many $i \in I$. Show that the map f from $\bigoplus_{i \in I} M_i$ to N sending $(m_i)_i$ to $\sum_i f_i(m_i)$ is an R-linear map.

2.3 Matrices

In this section we will discuss how to use matrices to study R-linear maps between free modules of finite rank. Throughout the section R denotes a ring and F denotes a field.

Matrices representing linear maps

An $n \times 1$ matrix is called a **column n-vector**. A $1 \times n$ matrix is called a **row n-vector**.

Let M be a free R-module with $\mathscr{B} = \{u_1, u_2, \ldots, u_n\}$ as a basis over R. For the discussion in this section, the order of the elements in the basis does matter. We will use $\beta = (u_1, u_2, \ldots, u_n)$ to denote an **ordered** basis for M over R.

For $x = \sum_{j=1}^{n} b_j u_j \in M$, we call $(b_1, b_2, \ldots, b_n)$ the **coordinates** of x with respect to the ordered basis β, and this row vector will be denoted as x_β^{row}. On the other hand, we will use x_β^{col} to denote the column vector consisting of the coordinates of x with respect to β. To be precise,

$$x_\beta^{\text{col}} = \begin{pmatrix} b_1 \\ \vdots \\ b_n \end{pmatrix}_{n \times 1}$$

when $x = \sum_{j=1}^{n} b_j u_j$.

In Example 1.1.9 we have defined the sum of two matrices with entries in a ring. We can also define the product of two such matrices as follows.

Definition 2.3.1. Let R be a ring. Let $A = (a_{ij})_{\ell \times m}$ and $B = (b_{ij})_{m \times n}$ be matrices over R. Define the **product** of A and B to be

$$AB = (c_{ij})_{\ell \times n}, \qquad \text{where } c_{ij} = \sum_{k=1}^{m} a_{ik} b_{kj}.$$

Lemma 2.3.2. *Let R be a ring. The following identities hold for matrices with entries in R.*

(i) $(A_{\ell \times m} B_{m \times n}) C_{n \times p} = A_{\ell \times m} (B_{m \times n} C_{n \times p})$.

(ii) $(A_{\ell \times m} + B_{\ell \times m}) C_{m \times n} = A_{\ell \times m} C_{m \times n} + B_{\ell \times m} C_{m \times n}$.

(iii) $A_{\ell \times m}(B_{m \times n} + C_{m \times n}) = A_{\ell \times m} B_{m \times n} + A_{\ell \times m} C_{m \times n}$.

We leave the proof of this lemma as a routine exercise. See Exercise 1.

Let M and N be free R-modules of finite rank. Let $\beta = (u_1, u_2, \ldots, u_n)$ and $\beta' = (v_1, v_2, \ldots, v_m)$ be their respective ordered bases over R. An R-linear map f from M to N is completely determined by the images $f(u_1), \ldots, f(u_n)$ by Theorem 2.1.11. Let

$$f(u_j) = \sum_{i=1}^{m} a_{ij} v_i, \qquad j = 1, \ldots, n.$$

If x is an element in M such that $x = \sum_{j=1}^{n} b_j u_j$, then

$$f(x) = \sum_{j=1}^{n} b_j f(u_j) = \sum_{j=1}^{n} \left(b_j \sum_{i=1}^{m} a_{ij} v_i \right) = \sum_{i=1}^{m} \left(\sum_{j=1}^{n} a_{ij} b_j \right) v_i.$$

If $f(x) = \sum_{i=1}^{m} c_i v_i$, then $c_i = \sum_{j=1}^{n} a_{ij} b_j$. This can be expressed by

$$\begin{pmatrix} c_1 \\ \vdots \\ c_m \end{pmatrix}_{m \times 1} = \begin{pmatrix} a_{11} & \cdots & a_{1n} \\ \vdots & \ddots & \vdots \\ a_{m1} & \cdots & a_{mn} \end{pmatrix}_{m \times n} \begin{pmatrix} b_1 \\ \vdots \\ b_n \end{pmatrix}_{n \times 1}.$$

The matrix $A = \left(a_{ij} \right)_{m \times n}$ is called **the matrix representing f with respect to the bases β and β' in column notation**. Observe that

$$f(x)_{\beta'}^{\text{col}} = A x_{\beta}^{\text{col}}, \qquad \text{where } A = \left(f(u_1)_{\beta'}^{\text{col}} \quad f(u_2)_{\beta'}^{\text{col}} \quad \cdots \quad f(u_n)_{\beta'}^{\text{col}} \right).$$

We may identify x with x_{β}^{col} via the isomorphism sending u_i of M to e_i of R^m. It is feasible to treat the free module M as R^m and N as R^n.

Obviously, one may also choose to use matrices to represent linear maps in **row notation**. We will use A^{tr} to denote the **transpose** of A given by

$$A^{\text{tr}} = \left(b_{ij} \right)_{n \times m}, \qquad \text{where } b_{ij} = a_{ji}.$$

In row notation, we should have

$$x_{\beta}^{\text{row}} = (b_1, \ldots, b_n) \qquad \text{and} \qquad f(x)_{\beta'}^{\text{row}} = (c_1, \ldots, c_m).$$

Note that A^{tr} is the matrix representing f with respect to the bases β and β' in row notation. We have that

$$f(x)_{\beta'}^{\mathrm{row}} = x_{\beta}^{\mathrm{row}} A^{\mathrm{tr}}, \qquad \text{where } A^{\mathrm{tr}} = \begin{pmatrix} f(u_1)_{\beta'}^{\mathrm{row}} \\ f(u_2)_{\beta'}^{\mathrm{row}} \\ \vdots \\ f(u_n)_{\beta'}^{\mathrm{row}} \end{pmatrix}$$

for any x in M.

To use the column notation or the row notation is only a matter of preference. For the rest of this textbook, we will write everything in column notation. The reader should be able to translate all the corresponding results in row notation.

Example 2.3.3. Let x be a variable over $\mathbb{R}$ and consider $\mathscr{P}_n \subseteq \mathbb{R}[x]$. Let T be the integration map in Example 2.1.4 with the domain and codomain restricted to $\mathscr{P}_n$ and $\mathscr{P}_{n+1}$ respectively. We have

$$T(1) = x;$$
$$T(x) = x^2/2;$$
$$T(x^2) = x^3/3;$$
$$\vdots \qquad \vdots$$
$$T(x^n) = x^{n+1}/(n+1).$$

The matrix of T with respect to the standard ordered bases $(1, x, \ldots, x^n)$ and $(1, x, \ldots, x^n, x^{n+1})$ of $\mathscr{P}_n$ and $\mathscr{P}_{n+1}$ respectively is

$$\begin{pmatrix} 0 & 0 & 0 & \cdots & 0 & 0 \\ 1 & 0 & 0 & \cdots & 0 & 0 \\ 0 & 1/2 & 0 & \cdots & 0 & 0 \\ 0 & 0 & 1/3 & \cdots & 0 & 0 \\ \vdots & \vdots & \vdots & \ddots & \vdots & \vdots \\ 0 & 0 & 0 & \cdots & 1/n & 0 \\ 0 & 0 & 0 & \cdots & 0 & 1/(n+1) \end{pmatrix}_{(n+2)\times(n+1)}.$$

Definition 2.3.4. The **identity matrix** $I_n \in M_n(R)$ is the $n \times n$ matrix with 1 on the diagonal and 0 elsewhere. In other words, $I_n = (\delta_{ij})_{n \times n}$ where $\delta_{ij} = 1$ if $i = j$ and $\delta_{ij} = 0$ otherwise.

A matrix $A \in M_n(R)$ is an **invertible** matrix if there is a (unique) matrix $B \in M_n(R)$ such that $AB = BA = I_n$. We say that B is the **inverse matrix** of A and we write $B = A^{-1}$.

Remarks. (1) Let $AB = BA = I$. Suppose C is another matrix such that $AC = CA = I$. Then $C = IC = (BA)C = B(AC) = BI = B$. Hence the inverse matrix of an invertible matrix is unique. It makes sense to denote the inverse of A by A^{-1}.

(2) Let $A \in M_{m \times n}(R)$ and $B \in M_{n \times m}(R)$. Since products of matrices are in general non-commutative, we should be aware that $AB = I_m$ does not necessarily imply that $BA = I_n$ (see Exercise 9). However, when A is a square matrix over a *commutative* ring, we will see that the situation is different in Chapter 3. See Theorem 3.4.9(b).

The following lemma gives a list of basic properties regarding invertible matrices. We leave it as an easy exercise. See Exercise 2.

Lemma 2.3.5. *Let A and B be invertible matrices in $M_n(R)$. The following statements are true.*

(a) *The identity matrix is invertible and $I_n^{-1} = I_n$.*

(b) *The matrix A^{-1} is invertible and $(A^{-1})^{-1} = A$.*

(c) *The matrix AB is invertible and $(AB)^{-1} = B^{-1}A^{-1}$.*

Proposition 2.3.6. *Let M, N and P be free R-modules with finite ordered bases β, β' and β'' respectively.*

(a) *The matrix representing an R-linear map $h \colon M \to M$ with respect to β and β is the identity matrix of size $|\beta|$ if and only if $h = 1_M$.*

(b) *Let $f \colon M \to N$ and $g \colon N \to P$ be linear maps. Suppose that A is the matrix representing f with respect to β and β' and B is the matrix representing g with respect to β' and β''. The matrix representing gf*

with respect to β and β'' is BA.

$$\overset{\displaystyle \overset{gf}{\overbrace{\hspace{2cm}}}}{\overset{BA}{}}$$

$$M \xrightarrow[A]{f} N \xrightarrow[B]{g} P$$
$$\beta \qquad\quad \beta' \qquad\quad \beta''$$

(c) *Assume that $|\beta| = |\beta'| < \infty$. Then f is an isomorphism if and only if A is an invertible matrix. In this case, A^{-1} is the matrix representing f^{-1} with respect to β' and β.*

Proof. Let's first set up the bases and matrices in question:

$$\beta = \{u_1, u_2, \ldots, u_n\};$$
$$\beta' = \{v_1, v_2, \ldots, v_m\};$$
$$\beta'' = \{w_1, w_2, \ldots, w_\ell\};$$
$$A = (a_{ij})_{m\times n} \quad\text{and}\quad B = (b_{ij})_{\ell\times m}.$$

Part (a) is trivial. We skip the proof.

(b) By assumption, we have

$$f(u_j) = \sum_{i=1}^{m} a_{ij}v_i, \qquad j = 1, 2, \ldots, n, \text{ and}$$

$$g(v_j) = \sum_{i=1}^{\ell} b_{ij}w_i, \qquad j = 1, 2, \ldots, m.$$

This implies that

$$gf(u_j) = g\left(\sum_{k=1}^{m} a_{kj}v_k\right) = \sum_{k=1}^{m} a_{kj}g(v_k)$$
$$= \sum_{k=1}^{m}\left[a_{kj}\left(\sum_{i=1}^{\ell} b_{ik}w_i\right)\right] = \sum_{i=1}^{\ell}\left(\sum_{k=1}^{m} a_{kj}b_{ik}\right)w_i$$

for $j = 1, 2, \ldots, n$. Thus the matrix representing gf with respect to β and β'' is $C = (c_{ij})_{\ell\times n}$ where $c_{ij} = \sum_{k=1}^{m} b_{ik}a_{kj}$. Hence $C = BA$ by definition.

(c) Assume that $|\beta| = |\beta'| = n$.

The "only if" part: Assume that f is an isomorphism. Let D be the matrix representing f^{-1} with respect to β' and β. Since $f^{-1}f = 1_M$, we have that $DA = I_n$ by (a) and (b). Similarly we have $AD = I_n$ since $ff^{-1} = 1_N$. Hence A is invertible and $D = A^{-1}$.

The "if" part: Let $g\colon N \to M$ be the linear map represented by A^{-1} with respect to β' and β. By (b), the matrix representing gf with respect to β and β is $A^{-1}A = I_n$. From (a) we have that $gf = 1_M$. Similarly we have that $fg = 1_N$. Thus f is an isomorphism and $g = f^{-1}$. $\qquad\square$

Definition 2.3.7. The matrix representing 1_M with respect to β and β' is called the **base change** matrix of M from β to β'.

We have the following immediate result about any base change matrix.

Corollary 2.3.8. *Let M be an R-module of finite rank. Any base change matrix of M is invertible.*

Proof. The result follows from Proposition 2.3.6(c) since 1_M is an isomorphism from M to M. In fact, the base change matrix from β' to β is the inverse of the base change matrix from β to β'. $\qquad\square$

The converse of Corollary 2.3.8 is also true. See Exercise 5 or Proposition 4.1.1.

Example 2.3.9. Let α be the standard ordered basis for R^3 over R and let
$$\beta = \big(f_1 = (1,0,0),\ f_2 = (1,1,0),\ f_3 = (1,1,1)\big).$$

(a) Verify that β is also a basis for R^3.

(b) Find the base change matrices (i) from β to α and (ii) from α to β.

(c) Express $(4, -5, 36) \in R^3$ as a linear combination of the base elements in β.

Solution. (a) Note that

(2.3.1) $\qquad e_1 = f_1, \quad e_2 = f_2 - f_1 \quad \text{and} \quad e_3 = f_3 - f_2.$

This shows that β generates R^3 over R. Let
$$af_1 + bf_2 + cf_3 = (0,0,0)$$

where a, b, $c \in R$. This gives a system of linear equations

$$\begin{cases} a + b + c = 0, \\ \quad\;\; b + c = 0, \\ \qquad\quad c = 0, \end{cases}$$

over R. It is easy to see that its only solution is the trivial solution. Hence β is linearly independent over R. We have shown that β is a basis over R.

(b) For case (i), note that

$$1_{R^3}(f_1) = f_1 = e_1,$$
$$1_{R^3}(f_2) = f_2 = e_1 + e_2,$$
$$1_{R^3}(f_3) = f_3 = e_1 + e_2 + e_3.$$

The base change matrix from β to α is

$$A = \begin{pmatrix} 1 & 1 & 1 \\ 0 & 1 & 1 \\ 0 & 0 & 1 \end{pmatrix}.$$

For case (ii), note that

$$1_{R^3}(e_1) = e_1 = f_1,$$
$$1_{R^3}(e_2) = e_2 = -f_1 + f_2,$$
$$1_{R^3}(e_3) = e_3 = -f_2 + f_3$$

from (2.3.1). The base change matrix from α to β is

$$B = \begin{pmatrix} 1 & -1 & 0 \\ 0 & 1 & -1 \\ 0 & 0 & 1 \end{pmatrix}.$$

Observe that the inverse of 1_{R^3} is itself. From Proposition 2.3.6(c), we know that A and B must be inverse to each other. We leave it to the reader to check that this is indeed the case by applying direct computation.

(c) The coordinate of $(4, -5, 36)$ with respect to the standard ordered basis α is itself. We may apply the base change matrix B from α to β to find the coordinate of $(4, -5, 36)$ with respect to β:

$$B \begin{pmatrix} 4 \\ -5 \\ 36 \end{pmatrix} = \begin{pmatrix} 1 & -1 & 0 \\ 0 & 1 & -1 \\ 0 & 0 & 1 \end{pmatrix} \begin{pmatrix} 4 \\ -5 \\ 36 \end{pmatrix} = \begin{pmatrix} 9 \\ -41 \\ 36 \end{pmatrix}.$$

This says that $(4, -5, 36) = 9f_1 - 41f_2 + 36f_3$.   ◇

Corollary 2.3.10. *Let M and N be free R-modules of finite rank. Let f be an R-linear map from M to N. Suppose β and β' are ordered bases for M while γ and γ' are ordered bases for N. Let A be the matrix for f with respect to β and γ. Let P be the base change matrix from β' to β and Q be the base change matrix from γ' to γ. The matrix for f with respect to β' and γ' is $Q^{-1}AP$.*

When working on a problem, we usually choose β and γ to be the standard bases so that the base change matrices P and Q are easier to find.

$$M \xrightarrow[\substack{P \\ \text{base change}}]{\mathbf{1}_M} M \xrightarrow[A]{f} N \xrightarrow[\substack{Q^{-1} \\ \text{base change}}]{\mathbf{1}_N} N$$

$$\beta' \qquad\qquad \beta \qquad\quad \gamma \qquad\qquad \gamma'$$

Proof. From Proposition 2.3.6(b), $Q^{-1}AP$ is the matrix representing

$$f = \mathbf{1}_N \circ f \circ \mathbf{1}_M$$

with respect to β' and γ'.   □

Definition 2.3.11. Let A and $B \in M_{m \times n}(R)$. We say A is **equivalent** to B if there are invertible matrices P and Q such that $B = Q^{-1}AP$.

In Exercise 10, the reader is asked to verify that the "equivalence" defined in Definition 2.3.11 is indeed an equivalence relation. Corollary 2.3.10 tells us that two equivalent matrices may represent the same linear map, albeit with respect to different ordered bases.

Column rank and row rank

An $m \times n$ matrix A over the field F may be viewed as

$$\begin{pmatrix} c_1 & c_2 & \dots & c_n \end{pmatrix}_{m \times n}$$

where $\mathbf{c}_j$ is a column m-vector for each j. We call the subspace generated by $\mathbf{c}_1$, $\mathbf{c}_2$, ..., $\mathbf{c}_n$ the **column space** of A, which is a subspace of F^m. We call the dimension of the column space of A the **column rank** of A, and we denote it by $\operatorname{col} \operatorname{rk} A$. Similarly, if we view A as

$$\begin{pmatrix} \mathbf{r}_1 \\ \mathbf{r}_2 \\ \dots \\ \mathbf{r}_m \end{pmatrix}_{m \times n}$$

where $\mathbf{r}_i$ is a row n-vector for each i. We call the subspace generated by $\mathbf{r}_1$, $\mathbf{r}_2$, ..., $\mathbf{r}_m$ the **row space** of A, which may be viewed as a subspace of F^n. The dimension of the row space of A is called the **row rank** of A and is denoted by $\operatorname{row} \operatorname{rk} A$.

Proposition 2.3.12. *Let V and W be finite dimensional vector spaces over F. If A is one of the matrix representing the linear transformation $T \colon V \to W$ in column notation, then $\operatorname{rk} T = \operatorname{col} \operatorname{rk} A$.*

Proof. Let $A = (a_{ij})_{m \times n}$ be the matrix representing T with respect to ordered bases β and γ. Let $\beta = (v_1, v_2, \ldots, v_n)$ and $\gamma = (w_1, w_2, \ldots, w_m)$. Let $U \colon W \to R^m$ be the linear map sending w_i to e_i. By Exercise 5, §2.1, we have that U is an isomorphism. Note that

$$\begin{aligned} (UT)(v_j) &= U(T(v_j)) = U\left(\sum_{i=1}^m a_{ij} w_i\right) = \sum_{i=1}^m a_{ij} e_i \\ &= \begin{pmatrix} a_{1j} \\ \vdots \\ a_{mj} \end{pmatrix} = T(u_j)_\gamma^{\mathrm{col}} \end{aligned}$$

if we assume the elements in F^m are column vectors. Hence the image of UT is the column space of A. We have that $\operatorname{col} \operatorname{rk} A = \operatorname{rk} UT = \operatorname{rk} T$ by Exercise 13, §2.2. $\qquad \square$

Exercises 2.3

For the following exercises, R denotes a ring and F denotes a field.

1. Prove Lemma 2.3.2.

2. Prove Lemma 2.3.5.

3. Show that $M_n(R)$ is a (usually non-commutative) ring.

4. Let $A \in M_n(R)$ such that $A = P_1 P_2 \ldots P_n$ where the P_i's are all invertible matrices in $M_n(R)$. Show that A is invertible in $M_n(R)$ and that $A^{-1} = P_n^{-1} \cdots P_2^{-1} P_1^{-1}$.

5. Let $\beta = (u_1, u_2, \ldots, u_n)$ be an ordered basis for a free R-module M of finite rank. Let $P = (p_{ij})_{n \times n}$ be an invertible matrix and let

$$v_j = \sum_{i=1}^{n} p_{ij} u_i, \qquad j = 1, 2, \ldots, n.$$

Show that $\beta' = (v_1, v_2, \ldots, v_n)$ is an ordered basis for M over R and show that the base change from β' to β is P.

6. Let D be the differential map from $\mathscr{P}_n$ to $\mathscr{P}_{n-1}$ in Example 2.1.4. Find the matrix representing f with respect to the standard ordered bases of $\mathscr{P}_n$ and $\mathscr{P}_{n-1}$.

7. Is the conjugation map f from $\mathbb{C}$ to $\mathbb{C}$ a $\mathbb{C}$-linear transformation? If yes, find the matrix representing f with respect to $\{1\}$ and $\{1\}$. Is f an $\mathbb{R}$-linear transformation? If yes, find the matrix representing f with respect to $(1, i)$ and $(1, i)$.

8. Let $P = \begin{pmatrix} 1 & 1 \\ 1 & 1 \end{pmatrix} \in M_2(R)$. Let T be the map from $M_2(R)$ to itself sending X to PX. Show that T is an R-linear map. Find the matrix representing T with respect to $\alpha = (e_{11}, e_{12}, e_{21}, e_{22})$ and α.

9. Find matrices A and B over $\mathbb{Z}$ such that AB is an identity matrix while BA is not!

10. Show that the "equivalence" defined in Definition 2.3.11 is indeed an equivalence relation.

11. Let M and N be R-modules with bases of size n and m respectively. Show that $\mathrm{Hom}_R(M, N) \cong M_{m \times n}(R)$.

12. Let $f\colon M_{m\times n}(R) \to M_{n\times m}(R)$ be the map sending A to A^{tr}. Is f an R-linear map? If yes, find the matrix representing f with respect to $\{e_{11}, e_{12}, e_{13}, e_{21}, e_{22}, e_{23}\}$ and $\{e_{11}, e_{12}, e_{21}, e_{22}, e_{31}, e_{32}\}$ when $m = 2$ and $n = 3$.

13. Let $\{x_1, x_2, \ldots, x_k\}$ and $\{y_1, y_2, \ldots, y_k\}$ be two linearly independent sets in an n-dimensional vector space V over F. Let β be an ordered basis for V over F. Let v_i and w_i be $(x_i)_\beta^{\mathrm{col}}$ and $(y_i)_\beta^{\mathrm{col}}$ respectively.

    (a) Show that $\{v_1, v_2, \ldots, v_k\}$ and $\{w_1, w_2, \ldots, w_k\}$ are two linearly independent sets in F^n.

    (b) Show that there is an invertible square matrix A with $Av_i = w_i$ for $i = 1, \ldots, k$.

14. Let $A \in M_{m\times n}(F)$. Let D and E be invertible matrices of size m and n respectively. Show that $\operatorname{col\,rk} DA = \operatorname{col\,rk} A = \operatorname{col\,rk} AE$.

15. In this exercise we will express a matrix in block form. Suppose all the matrices in this exercise have entries in the ring R. Show that

$$
\begin{pmatrix} A_{\ell\times m} & B_{\ell\times n} \\ C_{p\times m} & D_{p\times n} \end{pmatrix}_{(\ell+p)\times(m+n)} \begin{pmatrix} E_{m\times r} & F_{m\times s} \\ G_{n\times r} & H_{n\times s} \end{pmatrix}_{(m+n)\times(r+s)}
$$
$$
= \begin{pmatrix} AE + BG & AF + BH \\ CE + DG & CF + DH \end{pmatrix}_{(\ell+p)\times(r+s)}.
$$

This says that the product of matrices may be carried out in block form.

CHAPTER 3

Determinant

In this chapter we continue to study various properties of matrices. One important tool is the *determinant* of a matrix.

We first give the formal definition, and then we use the definition to derive all the "familiar" properties of the determinant. Basically speaking, we will discuss how an *elementary matrix operation* affects the determinant. We will also review the process of *Gaussian elimination*, which involves finding the *reduced row echelon form* of a matrix. Note that we no longer need to work over a field. We have to see how far we can go through the process when the scalars only come from elements of a ring. By doing so, we also get a deeper understanding of matrices in general. For example, we will obtain more results on invertible matrices, on the rank of a matrix, etc.

3.1 Basics of the determinant

In this section we will give the rigorous definition of the determinant. The usage of determinant is NOT restricted to matrices with entries in a field.

A short review of the symmetric group S_n

The symmetric group S_n consists of bijections from $\{1, 2, \ldots, n\}$ to itself. A bijection is also called a **permutation**.

Let $S = \{1, 2, 3, 4, 5\}$. A permutation $\sigma \in S_5$ given by

$$
\begin{array}{rccc}
\sigma : & S & \longrightarrow & S \\
& 1 & \longmapsto & 3 \\
& 2 & \longmapsto & 4 \\
& 3 & \longmapsto & 1 \\
& 4 & \longmapsto & 5 \\
& 5 & \longmapsto & 2
\end{array}
$$

may be expressed as

$$
\sigma = \begin{pmatrix} 1 & 2 & 3 & 4 & 5 \\ 3 & 4 & 1 & 5 & 2 \end{pmatrix}.
$$

The upper row places the elements in S and the second row gives their corresponding values. The numbers in the first row need not obey the usual order. For example we may also write

$$
\sigma = \begin{pmatrix} 1 & 4 & 5 & 3 & 2 \\ 3 & 5 & 2 & 1 & 4 \end{pmatrix}.
$$

There is a more efficient way to express elements in S_n called the **cycle notation**. An r-**cycle** is a series of r elements in $S = \{1, 2, 3, \ldots, n\}$ in the form

$$
\gamma = (n_1 \ n_2 \ n_3 \ \ldots \ n_r).
$$

The numbers n_1, n_2, ..., n_r must be distinct. The cycle γ expresses the permutation

$$
n_1 \xoverset{\gamma}{\longmapsto} n_2 \xoverset{\gamma}{\longmapsto} n_3 \xset{\gamma}{\longmapsto} \cdots \xoverset{\gamma}{\longmapsto} n_r \xoverset{\gamma}{\longmapsto} n_1
$$

while $\gamma(n) = n$ for $n \notin \{n_1, n_2, \ldots, n_r\}$. The cycle $(n_2\ n_3\ n_4\ \cdots\ n_r\ n_1)$ expresses the same permutation as the cycle $(n_1\ n_2\ n_3\ \ldots\ n_r)$. There are r different ways to express the same cycle.

A 1-cycle is simply the identity map on S. It may also be written as 1. The product of two cycles is the composite of the two permutations. An arbitrary permutation may be expressed as the product of several cycles. For example, let

$$\sigma = (1\ 2\ 3)(3\ 4)(5)(1\ 5\ 3) \in S_5.$$

The 1-cycle (5) above is redundant. One-cycles are usually omitted in a product of cycles. We have

$$
\begin{array}{ccc}
\sigma: & \{1,2,3,4,5\} & \longrightarrow & \{1,2,3,4,5\} \\
& 1 & \longmapsto & 5 \\
& 2 & \longmapsto & 3 \\
& 3 & \longmapsto & 2 \\
& 4 & \longmapsto & 1 \\
& 5 & \longmapsto & 4.
\end{array}
$$

We may also write $\sigma = (1\ 5\ 4)(2\ 3)$, which is a better expression.

Definition 3.1.1. We say two cycles are **disjoint** if they have no common entries.

Note that two disjoint cycles commute with each other.

Theorem 3.1.2. *Every permutation in S_n may be expressed as a product of disjoint cycles. The product is unique if we overlook the 1-cycles and disregard the different expressions of each cycle and the order of the disjoint cycles appearing in the product.*

We will omit the proof of the Theorem 3.1.2 because it is elementary, routine and tedious. The reader please refer to a textbook on modern algebra or on group theory.

Example 3.1.3. Let $\sigma = (1\ 3\ 5)(2\ 4) = (1\ 3\ 5)(6)(2\ 4) = (4\ 2)(5\ 1\ 3)$ be a permutation in S_6. The 3 expressions of σ as a product of disjoint cycles are deemed one and the same thing.

One drawback for the cycle notation is that one may not be sure which S_n the given product belongs in. For example, (1 2 3) may be seen as an element in S_n for any $n \geq 3$.

Definition 3.1.4. A 2-cycle is called a **transposition**.

Proposition 3.1.5. *Any permutation in S_n may be expressed as a product of transpositions.*

Proof. Note that $(n_1 \ n_2 \ \cdots \ n_r) = (n_1 \ n_r)(n_1 \ n_{r-1}) \cdots (n_1 \ n_2)$ is a product of $r - 1$ transpositions. In general, a permutation is a product of disjoint cycles by Theorem 3.1.2. We may decompose each cycle into a product of transpositions albeit losing the disjointness.   □

There may be many different ways to decompose a permutation into a product of transpositions. However, one thing remains the same as we can see in the following result.

Theorem 3.1.6. *Let $\sigma \in S_n$. If*

$$\sigma = \tau_1 \tau_2 \cdots \tau_s = \tau_1' \tau_2' \cdots \tau_t'$$

where τ_i and τ_j' are transpositions for all i and j. Then s and t are of the same parity.

Before we prove this result, we will first derive the following formulae.

Lemma 3.1.7. *Let u and v be non-negative integers. Let*

$$a, \ b, \ m_1, \ \ldots, \ m_u, \ n_1, \ \ldots, \ n_v$$

be distinct numbers in $\{1, 2, \ldots, n\}$. Then

   (i) $(a \ b)(a \ m_1 \ \ldots \ m_u \ b \ n_1 \ \ldots \ n_v) = (a \ m_1 \ \ldots \ m_u)(b \ n_1 \ \ldots \ n_v)$;

   (ii) $(a \ b)(a \ m_1 \ \ldots \ m_u)(b \ n_1 \ \ldots \ n_v) = (a \ m_1 \ \ldots \ m_u \ b \ n_1 \ \ldots \ n_v)$.

Proof. Without loss of generality we may assume that $u \leq v$ since the two disjoint cycles $(a \ m_1 \ \ldots \ m_u)$ and $(b \ n_1 \ \ldots \ n_v)$ commute with each other and in the long cycle, we may also exchange the order of the two sequences $a, m_1, \ldots, m_u$ and $b, n_1, \ldots, n_v$.

To prove (i), let $\sigma = (a\ m_1\ \ldots\ m_u\ b\ n_1\ \ldots\ n_v)$.

Case 1. Suppose $u = v = 0$. Then

$$(a\ b)\sigma = (a\ b)(a\ b) = 1 = (a)(b).$$

Case 2. Suppose $u = 0$ and $v = 1$. Then

$$(a\ b)\sigma = (a\ b)(a\ b\ n_1) = (b\ n_1) = (a)(b\ n_1).$$

Case 3. Suppose $u = 0$ and $v \geq 2$. Then

$$a \xrightarrow{\sigma} b \xrightarrow{(a\ b)} a$$
$$b \xrightarrow{\sigma} n_1 \xrightarrow{(a\ b)} n_1$$
$$n_i \xrightarrow{\sigma} n_{i+1} \xrightarrow{(a\ b)} n_{i+1}, \quad 1 \leq i \leq v - 1,$$
$$n_v \xrightarrow{\sigma} a \xrightarrow{(a\ b)} b.$$

Hence $(a\ b)\sigma = (a)(b\ n_1\ \ldots\ n_v)$.

Case 4. Suppose $u = v = 1$. Then

$$(a\ b)\sigma = (a\ b)(a\ m_1\ b\ n_1) = 1 = (a\ m_1)(b\ n_1).$$

Case 5. Suppose $u = 1$ and $v \geq 2$. Then

$$a \xrightarrow{\sigma} m_1 \xrightarrow{(a\ b)} m_1$$
$$m_1 \xrightarrow{\sigma} b \xrightarrow{(a\ b)} a$$
$$b \xrightarrow{\sigma} n_1 \xrightarrow{(a\ b)} n_1$$
$$n_i \xrightarrow{\sigma} n_{i+1} \xrightarrow{(a\ b)} n_{i+1}, \quad 1 \leq i \leq v - 1,$$
$$n_v \xrightarrow{\sigma} a \xrightarrow{(a\ b)} b.$$

Hence $(a\ b)\sigma = (a\ m_1)(b\ n_1\ \ldots\ n_v)$.

Case 6. Suppose $2 \leq u \leq v$. Then

$$a \xrightarrow{\sigma} m_1 \xrightarrow{(a\ b)} m_1$$
$$m_i \xrightarrow{\sigma} m_{i+1} \xrightarrow{(a\ b)} m_{i+1}, \quad 1 \leq i \leq u - 1,$$
$$m_u \xrightarrow{\sigma} b \xrightarrow{(a\ b)} a$$

$$b \xmapsto{\sigma} n_1 \xmapsto{(a\ b)} n_1$$

$$n_i \xmapsto{\sigma} n_{i+1} \xmapsto{(a\ b)} n_{i+1}, \quad 1 \le i \le v - 1,$$

$$n_v \xmapsto{\sigma} a \xmapsto{(a\ b)} b.$$

Hence $(a\ b)\sigma = (a\ m_1\ \ldots\ m_u)(b\ n_1\ \ldots\ n_v)$. The formula in (i) is established.

To prove (ii), note that

$$(a\ b)(a\ m_1\ \ldots\ m_u)(b\ n_1\ \ldots\ n_v)$$
$$= (a\ b)(a\ b)(a\ m_1\ \ldots\ m_u\ b\ n_1\ \ldots\ n_v), \quad \text{from (i)},$$
$$= (a\ m_1\ \ldots\ m_u\ b\ n_1\ \ldots\ n_v).$$

Hence the result.   □

Proof of Theorem 3.1.6. Let γ be an r-cycle. Define $N(\gamma) = r - 1$. Let $\sigma \in S_n$ and decompose

(3.1.1) $\sigma = \gamma_1 \gamma_2 \ldots \gamma_k$

into a product of disjoint cycles, where γ_i is an r_i-cycle for each i. Define

$$N(\sigma) = \sum_{i=1}^{k}(r_i - 1) = \sum_{i=1}^{k} N(\gamma_i).$$

The function N is well-defined by Theorem 3.1.2. The value $N(\sigma)$ will not be affected whether the decomposition in (3.1.1) includes 1-cycles or not. From the proof of Proposition 3.1.5, we actually know that σ may be expressed a product of $N(\sigma)$ transpositions. It suffices to show that t is always of the same parity as $N(\sigma)$ when $\sigma = \tau_1 \tau_2 \cdots \tau_s$ where the τ_i's are transpositions.

Let's first find out how multiplying a transposition on the left would affect N. Consider a 2-cycle $(a\ b)$. Suppose a and b lie in the same cycle in (3.1.1). We may assume that

$$\gamma_1 = (a\ m_1\ \ldots\ m_u\ b\ n_1\ \ldots\ n_v)$$

without loss of generality. Then

$$(a\ b)\sigma = (a\ b)(a\ m_1\ \ldots\ m_u\ b\ n_1\ \ldots\ n_v)\gamma_2 \cdots \gamma_k$$

$$= (a \; m_1 \; \ldots \; m_u)(b \; n_1 \; \ldots \; n_v)\gamma_2 \cdots \gamma_k$$

from Lemma 3.1.7. This is still a product of disjoint cycles. Hence

$$N((a \; b)\sigma) = u + v + \sum_{i=2}^{k} N(\gamma_i) = N(\gamma_1) - 1 + \sum_{i=2}^{k} N(\gamma_i) = N(\sigma) - 1.$$

On the other hand, if a and b lie in two different cycles in (3.1.1) (1-cycles included), we may assume

$$\gamma_1 = (a \; m_1 \; \ldots \; m_u) \quad \text{and} \quad \gamma_2 = (b \; n_1 \; \ldots \; n_v)$$

without loss of generality. Then

$$(a \; b)\sigma = (a \; b)(a \; m_1 \; \ldots \; m_u)(b \; n_1 \; \ldots \; n_v)\gamma_3 \cdots \gamma_k$$
$$= (a \; m_1 \; \ldots \; m_u \; b \; n_1 \; \ldots \; n_v)\gamma_3 \cdots \gamma_k.$$

We have that

$$N((a \; b)\sigma) = u + v + 1 + \sum_{i=3}^{k} N(\gamma_i)$$

$$= 1 + N(\gamma_1) + N(\gamma_2) + \sum_{i=3}^{k} N(\gamma_i) = N(\sigma) + 1.$$

In either case, we change the parity of N by multiplying a 2-cycle on the left. Hence the parity of $N(\tau_s\tau_{s-1}\cdots\tau_1\sigma)$ is changed s times from $N(\sigma)$. However,

$$N(\tau_s\tau_{s-1}\cdots\tau_1\sigma) = N(1) = 0.$$

This shows that s is odd if $N(\sigma)$ is odd, while s is even if $N(\sigma)$ is even. $\square$

From Theorem 3.1.6 we can make the following definition.

Definition 3.1.8. We say a permutation in S_n is an **even** permutation if it is a product of an even number of transpositions. Otherwise, we say it is an **odd** permutation. For $\sigma \in S_n$ we define the **sign** of σ to be

$$\text{sg}\,\sigma = \begin{cases} 1, & \text{if } \sigma \text{ is even;} \\ -1, & \text{if } \sigma \text{ is odd.} \end{cases}$$

We thus have the following immediate result.

Corollary 3.1.9. *For σ and $\pi \in S_n$ we have* $\mathrm{sg}(\sigma\pi) = \mathrm{sg}\,\sigma\,\mathrm{sg}\,\pi$.

An r-cycle is even if r is odd. An r-cycle is odd if r is even. We may use the corollary above to determine whether a permutation is even or odd.

The true definition of the determinant

Definition 3.1.10. Let $A = (a_{ij})$ be an $n \times n$ matrix with entries in a ring R. We define the **determinant** of A to be

$$\det A = \sum_{\sigma \in S_n} (\mathrm{sg}\,\sigma)\, a_{1\sigma(1)} a_{2\sigma(2)} \cdots a_{n\sigma(n)}.$$

Basically, each term in the summation consists of exactly one entry from each row and from each column. As one may also observe, the determinant only involves addition and multiplication. Thus it makes sense to discuss the determinant of a matrix over a ring.

Let $A_n = (a_{ij})_{n \times n}$. Let's express $\det A_n$ explicitly for small n's. Clearly, $\det A_1 = (\mathrm{sg}\,1) a_{11} = a_{11}$ and

$$\det A_2 = (\mathrm{sg}\,1) a_{11} a_{22} + \mathrm{sg}(1\,2) a_{12} a_{21} = a_{11} a_{22} - a_{12} a_{21}.$$

We also have

$$\det A_3 = (\mathrm{sg}\,1) a_{11} a_{22} a_{33} + \mathrm{sg}(1\,2\,3) a_{12} a_{23} a_{31} + \mathrm{sg}(1\,3\,2) a_{13} a_{21} a_{32}$$
$$+ \mathrm{sg}(2\,3) a_{11} a_{23} a_{32} + \mathrm{sg}(1\,3) a_{13} a_{22} a_{31} + \mathrm{sg}(1\,2) a_{12} a_{21} a_{33}$$
$$= a_{11} a_{22} a_{33} + a_{12} a_{23} a_{31} + a_{13} a_{21} a_{32}$$
$$- a_{11} a_{23} a_{32} - a_{13} a_{22} a_{31} - a_{12} a_{21} a_{33}.$$

If we only look at the three expressions above, we get the false impression that $\det A_4$ should be the sum of eight terms. In fact there are 24 terms in $\det A_4$. It takes too much space to express $\det A_4$ explicitly here. Let's just examine the sign of a few terms in the expression of $\det A_4$.

One might guess the sign of $a_{14} a_{21} a_{32} a_{43}$ in $\det A_4$ is positive by observing A_4:

$$\begin{pmatrix} & & & a_{14} \\ a_{21} & & & \\ & a_{32} & & \\ & & a_{43} & \end{pmatrix}.$$

However, in Definition 3.1.10 the σ corresponding to this term is

$$\sigma = \begin{pmatrix} 1 & 2 & 3 & 4 \\ 4 & 1 & 2 & 3 \end{pmatrix} = (1\ 4\ 3\ 2).$$

We should put a minus sign in front of the term $a_{14}a_{21}a_{32}a_{43}$.

To compute the sign of $a_{14}a_{23}a_{32}a_{41}$, observe the position of the entries:

$$\begin{pmatrix} & & & a_{14} \\ & & a_{23} & \\ & a_{32} & & \\ a_{41} & & & \end{pmatrix}.$$

Many students might guess the sign is negative from one's experience with $\det A_3$. The σ corresponding to this term is

$$\sigma = \begin{pmatrix} 1 & 2 & 3 & 4 \\ 4 & 3 & 2 & 1 \end{pmatrix} = (1\ 4)(2\ 3),$$

an even permutation. The sign of the term $a_{14}a_{23}a_{32}a_{41}$ should be positive.

How about the term $a_{14}a_{21}a_{33}a_{42}$:

$$\begin{pmatrix} & & & a_{14} \\ a_{21} & & & \\ & & a_{33} & \\ & a_{42} & & \end{pmatrix}?$$

The pattern of this term does not fit with any term in $\det A_3$. The σ corresponding to this term is

$$\sigma = \begin{pmatrix} 1 & 2 & 3 & 4 \\ 4 & 1 & 3 & 2 \end{pmatrix} = (1\ 4\ 2).$$

The sign of $a_{14}a_{21}a_{33}a_{42}$ is positive.

With enough patience, one may construct the entire summation for $\det A_4$. However, as n grows, the number of terms in $\det A_n$ grows even faster than the exponential growth. The number of terms in $\det A_n$ is $n!$. It is usually impossible to compute determinant with brutal force alone.

Next we give an alternative expression for $\det A$.

Lemma 3.1.11. *Let R be a ring and let $A = (a_{ij}) \in M_n(R)$. Then*

$$\det A = \sum_{\tau \in S_n} (\operatorname{sg} \tau) a_{\tau(1)1} a_{\tau(2)2} \cdots a_{\tau(n)n}.$$

In Definition 3.1.10, the entries in each summand for the determinant are ordered by the row. This lemma says that the entries in each summand may also be ordered by the column.

Proof. By Corollary 3.1.9, we have that $\operatorname{sg} \sigma = \operatorname{sg} \sigma^{-1} = \operatorname{sg} 1 = 1$. Thus $\operatorname{sg} \sigma = \operatorname{sg} \sigma^{-1}$ for any $\sigma \in S_n$. Using Definition 3.1.10 but reordering the a_{ij}'s by j, we have that

$$
\begin{aligned}
\det A &= \sum_{\sigma \in S_n} (\operatorname{sg} \sigma) \, a_{\sigma^{-1}(1)1} a_{\sigma^{-1}(2)2} \cdots a_{\sigma^{-1}(n)n} \\
&= \sum_{\tau \in S_n} (\operatorname{sg} \tau^{-1}) a_{\tau(1)1} a_{\tau(2)2} \cdots a_{\tau(n)n}, \qquad \text{denoting } \sigma^{-1} \text{ by } \tau, \\
&= \sum_{\tau \in S_n} (\operatorname{sg} \tau) a_{\tau(1)1} a_{\tau(2)2} \cdots a_{\tau(n)n}
\end{aligned}
$$

since $\operatorname{sg} \tau^{-1} = \operatorname{sg} \tau$. $\qquad \square$

The following result is also useful for our discussion on the determinant.

Lemma 3.1.12. *Suppose given a ring homomorphism φ form the ring R to the ring S. Let $A = (a_{ij})$ be a square matrix of size n over R. Then $\det \big(\varphi(a_{ij})\big) = \varphi(\det A)$.*

Proof. The argument is straightforward. We have that

$$
\begin{aligned}
\det \big(\varphi(a_{ij})\big) &= \sum_{\sigma \in S_n} (\operatorname{sg} \sigma) \, \varphi(a_{1\sigma(1)}) \varphi(a_{2\sigma(2)}) \cdots \varphi(a_{n\sigma(n)}) \\
&= \sum_{\sigma \in S_n} \varphi \big((\operatorname{sg} \sigma) \, a_{1\sigma(1)} a_{2\sigma(2)} \cdots a_{n\sigma(n)} \big) \\
&= \varphi \left(\sum_{\sigma \in S_n} (\operatorname{sg} \sigma) \, a_{1\sigma(1)} a_{2\sigma(2)} \cdots a_{n\sigma(n)} \right) \\
&= \varphi(\det A).
\end{aligned}
$$

Hence the result. $\qquad \square$

Basic properties of the determinant

Next, we verify some well-known properties regarding the determinant.

Proposition 3.1.13. *Let R be a ring and let $A \in M_n(R)$. The following statements are true.*

(a) *The determinant of A is equal to the determinant of its transpose.*

(b) *If A has only zero entries in one of its rows (or in one of its columns), its determinant is 0.*

(c) *Let $r \in R$ and let B be the matrix obtained by replacing one of A's rows (or columns) with r times the said row (or column respectively). Then $\det B = r \det A$.*

(d) *Let B be the matrix obtained by exchanging two rows (or two columns) of A. Then $\det B = -\det A$.*

(e) *If A has repeated rows (or columns), its determinant is 0.*

(f) *Suppose*

$$A = \begin{pmatrix} a_{11} & a_{12} & \cdots & a_{1n} \\ a_{21} & a_{22} & \cdots & a_{2n} \\ \vdots & \vdots & & \vdots \\ b_1+c_1 & b_2+c_2 & \cdots & b_n+c_n \\ \vdots & \vdots & & \vdots \\ a_{n1} & a_{n2} & \cdots & a_{nn} \end{pmatrix} \rightarrow \text{the } i_0\text{-th row }.$$

The determinant of A equals

$$\det \begin{pmatrix} a_{11} & a_{12} & \cdots & a_{1n} \\ a_{21} & a_{22} & \cdots & a_{2n} \\ \vdots & \vdots & & \vdots \\ b_1 & b_2 & \cdots & b_n \\ \vdots & \vdots & & \vdots \\ a_{n1} & a_{n2} & \cdots & a_{nn} \end{pmatrix} + \det \begin{pmatrix} a_{11} & a_{12} & \cdots & a_{1n} \\ a_{21} & a_{22} & \cdots & a_{2n} \\ \vdots & \vdots & \ddots & \vdots \\ c_1 & c_2 & \cdots & c_n \\ \vdots & \vdots & \ddots & \vdots \\ a_{n1} & a_{n2} & \cdots & a_{nn} \end{pmatrix}.$$

We will leave the statement of the column version to the reader.

(g) *Let B be the matrix obtained by adding a scalar multiple of the i_1-th row (or column) to the i_2-th row (or column respectively) of A. Then $\det B = \det A$.*

Proof. Let $A = (a_{ij}) \in M_n(R)$.

(a) Let $A^{\mathrm{tr}} = (b_{ij})$. Then $b_{ij} = a_{ji}$ for all i and j. From definition we have that

$$
\begin{aligned}
\det A^{\mathrm{tr}} &= \sum_{\sigma \in S_n} (\mathrm{sg}\,\sigma)\, b_{1\sigma(1)} b_{2\sigma(2)} \cdots b_{n\sigma(n)} \\
&= \sum_{\sigma \in S_n} (\mathrm{sg}\,\sigma)\, a_{\sigma(1)1} a_{\sigma(2)2} \cdots a_{\sigma(n)n} = \det A
\end{aligned}
$$

from Lemma 3.1.11.

From (a), for the rest of the proof it suffices to prove the row version of (b)–(g) because any statement regarding the columns of a matrix applies to the corresponding rows of its transpose.

(b) Suppose the entries of the i_0-th row of A are all 0. We have $a_{i_0 j} = 0$ for all j. Then

$$
a_{1\sigma(1)} a_{2\sigma(2)} \cdots a_{i_0 \sigma(i_0)} \cdots a_{n\sigma(n)} = 0
$$

for all $\sigma \in S_n$. This implies that $\det A = 0$

(c) Let $B = (b_{ij})$ where

$$
b_{ij} = \begin{cases} r a_{ij}, & \text{if } i = i_0; \\ a_{ij}, & \text{otherwise.} \end{cases}
$$

Thus

$$
\begin{aligned}
\det B &= \sum_{\sigma \in S_n} (\mathrm{sg}\,\sigma)\, b_{1\sigma(1)} b_{2\sigma(2)} \cdots b_{n\sigma(n)} \\
&= \sum_{\sigma \in S_n} (\mathrm{sg}\,\sigma)\, r a_{1\sigma(1)} a_{2\sigma(2)} \cdots a_{n\sigma(n)} = r \det A.
\end{aligned}
$$

(d) Let B be the matrix obtained by exchanging the i_1-th row and the i_2-th row $(i_1 < i_2)$ of A. Let $B = (b_{ij})$. Then

$$
b_{ij} = \begin{cases} a_{i_2 j}, & \text{if } i = i_1; \\ a_{i_1 j}, & \text{if } i = i_2; \\ a_{ij}, & \text{otherwise.} \end{cases}
$$

We have

$$\det B = \sum_{\sigma \in S_n} (\text{sg } \sigma)\, b_{1\sigma(1)} b_{2\sigma(2)} \cdots b_{n\sigma(n)}$$

$$= \sum_{\sigma \in S_n} (\text{sg } \sigma)\, a_{1\sigma(1)} a_{2\sigma(2)} \cdots a_{i_2 \sigma(i_1)} \cdots a_{i_1 \sigma(i_2)} \cdots a_{n\sigma(n)}.$$

Consider $\pi = \sigma(i_1\ i_2)$. Then

$$\pi(i) = \begin{cases} \sigma(i_2), & \text{if } i = i_1; \\ \sigma(i_1), & \text{if } i = i_2; \\ \sigma(i), & \text{otherwise.} \end{cases}$$

We have

$$\det B = \sum_{\sigma \in S_n} (\text{sg } \sigma)\, a_{1\pi(1)} a_{2\pi(2)} \cdots a_{i_1 \pi(i_1)} \cdots a_{i_2 \pi(i_2)} \cdots a_{n\pi(n)}$$

$$= - \sum_{\pi \in S_n} (\text{sg } \pi)\, a_{1\pi(1)} a_{2\pi(2)} \cdots a_{i_1 \pi(i_1)} \cdots a_{i_2 \pi(i_2)} \cdots a_{n\pi(n)}$$

$$= - \det A$$

since $\text{sg } \sigma = - \text{sg } \pi$ for all $\sigma \in S_n$.

(e) Let x_{ij}, $i, j = 1, \ldots, n$, be variables over $\mathbb{Z}$. Consider the ring $S = \mathbb{Z}[\, x_{ij} : i, j = 1, \ldots, n\,]$, the polynomial ring of n^2 variables over $\mathbb{Z}$. Let $i_1 < i_2$ be two numbers between 1 and n. Consider the $n \times n$ matrix $X = (f_{ij})_{m \times n}$ where

$$f_{ij} = \begin{cases} x_{ij}, & \text{if } i \neq i_2; \\ x_{i_1, j}, & \text{if } i = i_2. \end{cases}$$

In other words, the i_1-th row and the i_2-th row are the same as in X. If we exchange these two rows of X, the matrix X is unchanged. Hence $\det X = - \det X$ from (d). We have $2 \det X = 0$. Since S is an integral domain of characteristic 0, we conclude that $\det X = 0$.

Now suppose in the matrix A, the i_1-th row and the i_2-th row are the same. Consider the ring homomorphism $\varphi \colon S \to R$ sending x_{ij} to a_{ij} for all i and j. Then

$$\det A = \det\big(\varphi(f_{ij})\big) = \varphi(\det X) = \varphi(0) = 0$$

by Lemma 3.1.12.

(f) This follows from the fact that

$$a_{1\sigma(1)}a_{2\sigma(2)}\cdots a_{n\sigma(n)}$$

$$= a_{1\sigma(1)}a_{2\sigma(2)}\cdots a_{i_0-1,\sigma(i_0-1)}\big(b_{\sigma(i_0)}+c_{\sigma(i_0)}\big)a_{i_0+1,\sigma(i_0+1)}\cdots a_{n\sigma(n)}$$

$$= a_{1\sigma(1)}a_{2\sigma(2)}\cdots a_{i_0-1,\sigma(i_0-1)}b_{\sigma(i_0)}a_{i_0+1,\sigma(i_0+1)}\cdots a_{n\sigma(n)}$$

$$+\, a_{1\sigma(1)}a_{2\sigma(2)}\cdots a_{i_0-1,\sigma(i_0-1)}c_{\sigma(i_0)}a_{i_0+1,\sigma(i_0+1)}\cdots a_{n\sigma(n)}.$$

(g) From (f) and (c) we have that $\det B$ equals the sum of $\det A$ and a scalar multiple of the determinant of a matrix with repeated rows. The latter determinant is 0 from (e). Thus $\det B = \det A$. □

Proposition 3.1.13 provides a more efficient method to compute the determinant.

Example 3.1.14. Let's try to find the determinant of

$$A = \begin{pmatrix} 2 & 0 & 0 & 1 \\ 0 & 1 & 3 & -3 \\ -2 & -3 & -5 & 2 \\ 4 & -4 & 4 & -6 \end{pmatrix}$$

without resorting to the definition. We have that $\det A$ is equal to

$$2\det\begin{pmatrix} 1 & 0 & 0 & 1 \\ 0 & 1 & 3 & -3 \\ -1 & -3 & -5 & 2 \\ 2 & -4 & 4 & -6 \end{pmatrix}, \quad \text{factoring from col 1,}$$

$$= 4\det\begin{pmatrix} 1 & 0 & 0 & 1 \\ 0 & 1 & 3 & -3 \\ -1 & -3 & -5 & 2 \\ 1 & -2 & 2 & -3 \end{pmatrix}, \quad \text{factoring from row 4,}$$

$$= 4\det\begin{pmatrix} 1 & 0 & 0 & 0 \\ 0 & 1 & 3 & -3 \\ -1 & -3 & -5 & 3 \\ 1 & -2 & 2 & -4 \end{pmatrix}, \quad \text{adding } (-1)(\text{col 1}) \text{ to col 4}$$

$$= 4\det \begin{pmatrix} 1 & 0 & 0 & 0 \\ 0 & 1 & 0 & 0 \\ -1 & -3 & 4 & -6 \\ 1 & -2 & 8 & -10 \end{pmatrix}, \qquad \begin{array}{l} \text{adding } (-3)(\text{col } 2) \text{ to col } 3 \\ \text{adding } 3(\text{col } 2) \text{ to col } 4 \end{array}$$

$$= 32\det \begin{pmatrix} 1 & 0 & 0 & 0 \\ 0 & 1 & 0 & 0 \\ -1 & -3 & 1 & -3 \\ 1 & -2 & 2 & -5 \end{pmatrix}, \quad \text{factoring from col 3 and col 4,}$$

$$= 32\det \begin{pmatrix} 1 & 0 & 0 & 0 \\ 0 & 1 & 0 & 0 \\ -1 & -3 & 1 & 0 \\ 1 & -2 & 2 & 1 \end{pmatrix}, \quad \text{adding } 3(\text{col } 3) \text{ to col } 4$$

$$= 32, \quad \text{by Exercise 5, §3.1.}$$

Exercises 3.1

Throughout these exercises, R denotes a ring.

1. Show that $S_n = \langle (1\ 2), (1\ 3), \ldots, (1\ n) \rangle$ as a group.

2. Find sg$(1\ 2\ 3\ 4\ 5)(2\ 3\ 4\ 5)(1\ 3\ 5\ 7\ 9)(2\ 5\ 9)$.

3. Find the determinant of

$$A = e_{11} + 2e_{13} + 2e_{22} + 4e_{24} + 3e_{31} + 4e_{33} + e_{42} + e_{44} \in M_4(\mathbb{Z}_6)$$

using Definition 3.1.10 directly.

Let $A = (a_{ij}) \in M_{m \times n}(R)$. We call the a_{ii}'s the **diagonal** entries of A. If the entries of a matrix are all zero except for the diagonal entries, it is called a **diagonal** matrix. The matrix A is called an **upper triangular** matrix if $a_{ij} = 0$ for $i > j$. It is called a **lower triangular** matrix if $a_{ij} = 0$ for $i < j$. Together they are called **triangular** matrices.

4. Let A be a square diagonal matrix. Show that $\det A$ equals the product of the diagonal entries. In particular, the determinant of the identity matrix is 1.

Figure 3.1: Suppose the shaded area contains all the nonzero entries. The left-hand side indicates a lower triangular matrix. The right-hand side indicates an upper triangular matrix.

5. Let A be a square triangular matrix. Show that $\det A$ equals the product of the diagonal entries.

Let $A = (a_{ij})$ be a square matrix of size n. The "diagonal" of A which is traversed in the "northeast" direction (the entries from a_{n1} to a_{1n}) is called the **skew diagonal** of A. The matrix A is called an **upper anti-triangular** matrix if the entries below the skew diagonal are all zero. It is called a **lower anti-triangular** if the entries above the skew diagonal are all zero. Together they are called **anti-triangular** matrices.

6. Show that the determinant of a square anti-triangular matrix of size n equals $(-1)^{\lfloor n/2 \rfloor}$ times the product of its skew diagonal entries. Here $[x]$ is the **floor function** whose value is the greatest integer less than or equal to x.

7. Prove Proposition 3.1.13(e) directly using the definition of determinant. (Hint: Suppose the i_1-th row and the i_2-th row are repeated.

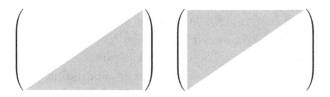

Figure 3.2: Suppose the shaded area contains all the nonzero entries. The left-hand side indicates a lower anti-triangular matrix. The right-hand side indicates an upper anti-triangular matrix.

Divide the sum into two parts, the summation over A_n and the summation over $\sigma(i_1\ i_2)$ for $\sigma \in A_n$. Here A_n is the **alternating** n-group, the group consisting of all even permutation in S_n.)

8. A matrix of the form

(3.1.2)
$$
\begin{pmatrix}
1 & a_1 & a_1^2 & \cdots & a_1^{n-1} \\
1 & a_2 & a_2^2 & \cdots & a_2^{n-1} \\
\multicolumn{5}{c}{\dotfill} \\
\multicolumn{5}{c}{\dotfill} \\
1 & a_n & a_n^2 & \cdots & a_n^{n-1}
\end{pmatrix}, \qquad a_1, a_2, \ldots, a_n \in R,
$$

is called a **Vandermonde matrix**. Show that the determinant of the Vandermonde matrix in (3.1.2) is $\displaystyle\prod_{1 \le i < j \le n} (a_j - a_i)$.

3.2 Matrix operations

Elementary row- and column operations are integral to the study of matrices over a field. In this section we discuss row and column operations in general and their consequences.

Throughout this section, R denotes a ring.

Row operations and column operations

Before we start let's first make an observation. Let $A \in M_{m \times n}(R)$. For our purpose we will view A as a column m-vector with row n-vectors as its entries. In other words, we may write $A = \begin{pmatrix} \mathbf{r}_1 \\ \mathbf{r}_2 \\ \vdots \\ \mathbf{r}_m \end{pmatrix}$ where $\mathbf{r}_i$ is a row n-vector for each i. Notice that

$$
\begin{pmatrix} a_1, & a_2, & \ldots, & a_m \end{pmatrix}
\begin{pmatrix} \mathbf{r}_1 \\ \mathbf{r}_2 \\ \vdots \\ \mathbf{r}_m \end{pmatrix}
= a_1\mathbf{r}_1 + a_2\mathbf{r}_2 + \cdots + a_m\mathbf{r}_m.
$$

If we multiply A by a row m-vector from the left, we obtain a linear combination of the rows of A. In general

$$
\begin{pmatrix}
a_{11} & a_{12} & \cdots & a_{1m} \\
a_{21} & a_{22} & \cdots & a_{2m} \\
\vdots & \vdots & \ddots & \vdots \\
a_{\ell 1} & a_{\ell 2} & \cdots & a_{\ell m}
\end{pmatrix}
A =
\begin{pmatrix}
a_{11}\mathbf{r}_1 + a_{12}\mathbf{r}_2 + \cdots + a_{1m}\mathbf{r}_m \\
a_{21}\mathbf{r}_1 + a_{22}\mathbf{r}_2 + \cdots + a_{2m}\mathbf{r}_m \\
\vdots \\
a_{\ell 1}\mathbf{r}_1 + a_{\ell 2}\mathbf{r}_2 + \cdots + a_{\ell m}\mathbf{r}_m
\end{pmatrix}.
$$

If we multiply A by an $\ell \times m$ matrix from the left, we obtain an $\ell \times n$ matrix whose rows are linear combinations of the rows of A. Thus we may say that to multiply A by a matrix from the left is equivalent to performing a **row operation** on A.

Similarly, we may view A as a row n-vector with column m-vectors as its entries:

$$
A = \begin{pmatrix} \mathbf{c}_1, & \mathbf{c}_2, & \ldots, & \mathbf{c}_n \end{pmatrix}.
$$

We have that

$$
\begin{pmatrix} \mathbf{c}_1 & \mathbf{c}_2 & \cdots & \mathbf{c}_n \end{pmatrix}
\begin{pmatrix}
b_{11} & b_{12} & \cdots & b_{1p} \\
b_{21} & b_{22} & \cdots & b_{2p} \\
\vdots & \vdots & \ddots & \vdots \\
b_{n1} & b_{n2} & \cdots & b_{np}
\end{pmatrix}
$$

$$
= (b_{11}\mathbf{c}_1 + b_{21}\mathbf{c}_2 + \cdots + b_{n1}\mathbf{c}_n,\ b_{12}\mathbf{c}_1 + b_{22}\mathbf{c}_2
$$
$$
+ \cdots + b_{n2}\mathbf{c}_n, \ldots,\ b_{1p}\mathbf{c}_1 + b_{2p}\mathbf{c}_2 + \cdots + b_{np}\mathbf{c}_n).
$$

When we multiply A by an $n \times p$ matrix from the right, we obtain an $m \times p$ matrix whose columns are linear combinations of the columns of A. We may say that to multiply A by a matrix from the right is equivalent to performing a **column operation** on A.

Example 3.2.1. Let $\mathbf{r}_1 = (1, 2)$ and $\mathbf{r}_2 = (3, 4)$ be two row vectors. Make up two new rows from $\mathbf{r}_1$ and $\mathbf{r}_2$:

$$
2\mathbf{r}_1 + \mathbf{r}_2 = (5, 8);
$$
$$
\mathbf{r}_1 - 3\mathbf{r}_2 = (-8, -10).
$$

When we want to stack these two rows together, we may describe the process by the following identity:

$$\begin{pmatrix} 2 & 1 \\ 1 & -3 \end{pmatrix} \begin{pmatrix} 1 & 2 \\ 3 & 4 \end{pmatrix} = \begin{pmatrix} 5 & 8 \\ -8 & -10 \end{pmatrix}.$$

If we stack $\mathbf{r}_1$ on top of $\mathbf{r}_2$, we obtain the matrix in the middle. The matrix on the left tells us that the resulting matrix will be of two rows, and the rows of this matrix tell us what to do to obtain the new rows.

Elementary operations and elementary matrices

Among all row and column operations some are particularly important.

Let $A \in M_{m \times n}(R)$. Any one of the following three operations on the rows [columns] of A is called an **elementary row [column] operation**:

- Exchanging two rows [columns] of A;

- Multiplying a row [column] of A by a unit scalar;

- Adding a scalar multiple of a row [column] of A to another row [column].

These elementary operations are said to be of **type I, type II** or **type III** respectively.

An elementary row [column] operation can be induced by multiplying the target matrix by an appropriate matrix from the left [right]. The "appropriate" matrices are called **elementary matrices**. The elementary matrix is said to be of **type I, type II** or **type III** according to whether the elementary operation performed is of type I, type II or type III.

How do we find the elementary matrices? To perform an elementary row operation on A is equivalent to multiply by an elementary matrix E from the left. The resulting matrix is EA. How do we find E? We only need to perform the said elementary operation on I_m and we obtain $EI_m = E$. Similarly if we are to obtain the matrix of an elementary column operation we simply perform the same elementary column operation on I_n.

Next we find the elementary matrices explicitly. All matrices involved are assumed to be square matrices of appropriate sizes.

Elementary Matrices of Type I. If we are to exchange the i-th and the j-th rows of $A_{n \times n}$, it needs to be multiplied by the matrix

$$
P_{ij}^m \;=\; \begin{array}{c} \\ \\ i \to \\ \\ j \to \\ \\ \\ \\ \end{array}
\begin{pmatrix}
1 & & & & & & & & \\
& 1 & & & & & & & \\
& & \ddots & & & & & & \\
& & & 0 & \cdots & 1 & & & \\
& & & \vdots & & \vdots & & & \\
& & & 1 & \cdots & 0 & & & \\
& & & & & & 1 & & \\
& & & & & & & \ddots & \\
& & & & & & & & 1 \\
\end{pmatrix}
$$

$$
= I_m - e_{ii} - e_{jj} + e_{ij} + e_{ji}
$$

from the left. Here, the e_{ij}'s are all assumed to be of size m. If instead we multiply A by P_{ij}^n from the right, we exchange the i-th and the j-th column of A. When the size of the elementary matrix is understood, we may simply write P_{ij} instead. This remark applies to the elementary matrices of types II and III as well.

Elementary Matrices of Type II. If we are to multiply the i-th row of A by a unit scalar u, we need to multiply A by the matrix

$$
D_i^m(u) = \begin{array}{c} \\ \\ \\ i \to \\ \\ \\ \end{array}
\begin{pmatrix}
1 & & & & & \\
& 1 & & & & \\
& & \ddots & & & \\
& & & u & & \\
& & & & \ddots & \\
& & & & & 1 \\
\end{pmatrix}
= I_m + (u-1)e_{ii}
$$

from the left. If instead we multiply A by $D_i^n(u)$ from the right, we are multiplying the i-th column of A by the unit scalar u.

Elementary Matrices of Type III. If we want to add b times the j-th row to the i-th row in A, we need to multiply A by the matrix

$$T_{ij}^m(b) = \begin{array}{c} \\ \\ i \rightarrow \\ \\ j \rightarrow \\ \\ \end{array} \begin{pmatrix} 1 & & & & & & \\ & \ddots & & & & & \\ & & 1 & \cdots & b & & \\ & & & \ddots & \vdots & & \\ & & & & 1 & & \\ & & & & & \ddots & \\ & & & & & & 1 \end{pmatrix} = I_m + be_{ij}$$

from the left. If instead we multiply A by $T_{ij}^n(b)$ from the right we obtain a matrix obtained by adding b times the i-th column to j-th column in A.

Example 3.2.2. Compute the following products.

(a) $\begin{pmatrix} 0 & 0 & 1 \\ 0 & 1 & 0 \\ 1 & 0 & 0 \end{pmatrix} \begin{pmatrix} 1 & 2 \\ -7 & 2 \\ -4 & 9 \end{pmatrix}$

(b) $\begin{pmatrix} 1 & 2 \\ -7 & 2 \\ -4 & 9 \end{pmatrix} \begin{pmatrix} 1 & 0 \\ 0 & -0.9 \end{pmatrix}$

(c) $\begin{pmatrix} 1 & 0 & 0 \\ 0 & 1 & 0 \\ 7 & 0 & 1 \end{pmatrix} \begin{pmatrix} 1 & 2 \\ -7 & 2 \\ -4 & 9 \end{pmatrix}$

(d) $\begin{pmatrix} 1 & 0 & 0 \\ 0 & -0.9 & 0 \\ 0 & 0 & 1 \end{pmatrix} \begin{pmatrix} 1 & 0 & 0 \\ 0 & 1 & 0 \\ 7 & 0 & 1 \end{pmatrix} \begin{pmatrix} 1 & 2 \\ -7 & 2 \\ -4 & 9 \end{pmatrix} \begin{pmatrix} 0 & 1 \\ 1 & 0 \end{pmatrix}$

(e) $\begin{pmatrix} 1 & 2 & 1 \\ -7 & 2 & 3 \\ -4 & 9 & 7 \end{pmatrix} \begin{pmatrix} 1 & 0 & 0 \\ 0 & 3 & 0 \\ 7 & 0 & 1 \end{pmatrix}$

Solution. (a) The matrix on the left is an elementary matrix of type I. We need to exchange the first and the third row of the matrix on the right. Hence the product is

$$\begin{pmatrix} 0 & 0 & 1 \\ 0 & 1 & 0 \\ 1 & 0 & 0 \end{pmatrix} \begin{pmatrix} 1 & 2 \\ -7 & 2 \\ -4 & 9 \end{pmatrix} = \begin{pmatrix} -4 & 9 \\ -7 & 2 \\ 1 & 2 \end{pmatrix}.$$

(b) The matrix on the right is an elementary matrix of type II. We need to multiply the second column by -0.9. Thus

$$\begin{pmatrix} 1 & 2 \\ -7 & 2 \\ -4 & 9 \end{pmatrix} \begin{pmatrix} 1 & 0 \\ 0 & -0.9 \end{pmatrix} = \begin{pmatrix} 1 & -1.8 \\ -7 & -1.8 \\ -4 & -8.1 \end{pmatrix}.$$

(c) Here we add 7 times the first row to the third row and we have

$$\begin{pmatrix} 1 & 0 & 0 \\ 0 & 1 & 0 \\ 7 & 0 & 1 \end{pmatrix} \begin{pmatrix} 1 & 2 \\ -7 & 2 \\ -4 & 9 \end{pmatrix} = \begin{pmatrix} 1 & 2 \\ -7 & 2 \\ 3 & 23 \end{pmatrix}.$$

(d) We will perform two consecutive row operations, followed by an elementary column operation of type I. We have

$$\begin{pmatrix} 1 & 0 & 0 \\ 0 & -0.9 & 0 \\ 0 & 0 & 1 \end{pmatrix} \begin{pmatrix} 1 & 0 & 0 \\ 0 & 1 & 0 \\ 7 & 0 & 1 \end{pmatrix} \begin{pmatrix} 1 & 2 \\ -7 & 2 \\ -4 & 9 \end{pmatrix} \begin{pmatrix} 0 & 1 \\ 1 & 0 \end{pmatrix}$$

$$= \begin{pmatrix} 1 & 0 & 0 \\ 0 & -0.9 & 0 \\ 0 & 0 & 1 \end{pmatrix} \begin{pmatrix} 1 & 2 \\ -7 & 2 \\ 3 & 23 \end{pmatrix} \begin{pmatrix} 0 & 1 \\ 1 & 0 \end{pmatrix}$$

$$= \begin{pmatrix} 1 & 2 \\ 6.3 & -1.8 \\ 3 & 23 \end{pmatrix} \begin{pmatrix} 0 & 1 \\ 1 & 0 \end{pmatrix}$$

$$= \begin{pmatrix} 2 & 1 \\ -1.8 & 6.3 \\ 23 & 3 \end{pmatrix}.$$

We would have reached the same conclusion if we had performed the column operation followed by the two consecutive row operations.

(e) The matrix on the right is not an elementary matrix. However it is easy to see that it is obtained by performing two elementary column operations. We add 7 times the third column to the first column and multiply the second column by 3. We have

$$
\begin{pmatrix} 1 & 2 & 1 \\ -7 & 2 & 3 \\ -4 & 9 & 7 \end{pmatrix} \begin{pmatrix} 1 & 0 & 0 \\ 0 & 3 & 0 \\ 7 & 0 & 1 \end{pmatrix} = \begin{pmatrix} 8 & 6 & 1 \\ 14 & 6 & 3 \\ 45 & 27 & 7 \end{pmatrix}.
$$

There is no need to actually carry out the tedious process of multiplication. It is sufficient to view matrices as matrix operations.   ◇

Note that each of the elementary row [column] operations can be reversed by an elementary row [column] operation. This means that the elementary matrices must be invertible.

If we want to reverse the elementary row operation of exchanging the i-th and the j-th row of A, we simply exchange it back. It is possible that

$$
P_{ij}^{-1} = P_{ij}.
$$

If we want to reverse the elementary row operation of multiplying the i-th row by a unit u, we simply multiply the i-th row by u^{-1}. We should have

$$
D_i(u)^{-1} = D_i(u^{-1}).
$$

If we want to reverse the elementary row operation of adding b times the j-th row to the i-th row of A, we add $-b$ times the j-th row of A to the i-th row. It should follow that

$$
T_{ij}(b)^{-1} = T_{ij}(-b).
$$

We leave it as an exercise (see Exercise 1) to verify by straightforward calculation that all the conjectures above are true. We summarize the results in the following proposition.

Proposition 3.2.3. *Elementary matrices are invertible, and the inverse of an elementary matrix is an elementary matrix of the same type.*

Now let's find the determinant of the elementary matrices.

Lemma 3.2.4. *Let $b \in R$ and u be a unit of R. Suppose that P_{ij}, $D_i(u)$ and $T_{ij}(b)$ are elementary matrices of size n. Then*

(i) $\det P_{ij} = -1$;

(ii) $\det D_i(u) = u$;

(iii) $\det T_{ij}(b) = 1$.

Hence the determinant of the elementary matrices are all units in R.

Proof. (i) The elementary matrix P_{ij} is obtained by exchanging two rows of the identity matrix. Note that the determinant of the identity matrix is 1 by Exercise 4, §3.1. The result follows from Proposition 3.1.13(d).

(ii) The result follows from Proposition 3.1.13(c) since $D_i(u)$ is obtained by multiplying the i-th row of the identity matrix by u.

(iii) The matrix $T_{ij}(b)$ is obtained by adding b times the j-th row of the identity matrix to the i-th row. The result now follows from Proposition 3.1.13(g). □

Corollary 3.2.5. *Let A and E be square matrices of size n over a ring R and E be an elementary matrix. Then $\det(EA) = \det E \det A = \det(AE)$.*

Proof. The result follows from Proposition 3.1.13(c), (d) and (g) together with Lemma 3.2.4. □

Corollary 3.2.6. *Let R be a ring, $b \in R$ and u be a unit of R. Let P_{ij}, $D_i(u)$ and $T_{ij}(b)$ be elementary matrices of size n. Then*

(i) $\det P_{ij}^{-1} = -1$;

(ii) $\det D_i(u)^{-1} = u^{-1}$;

(iii) $\det T_{ij}(b)^{-1} = 1$.

Proof. Let E be an elementary matrix. Then

$$\det E \det E^{-1} = \det EE^{-1} = \det I_n = 1$$

from Corollary 3.2.5. The result follows from Lemma 3.2.4. The result also follows from Exercise 1 and Lemma 3.2.4. □

Exercises 3.2

Suppose all the matrices in the following exercises are over the ring R.

1. Let $b \in R$ and u be a unit of R. Verify the following identities by straightforward calculation.

   (a) $P_{ij}^{-1} = P_{ij}$.

   (b) $D_i(u)^{-1} = D_i(u^{-1})$.

   (c) $T_{ij}(b)^{-1} = T_{ij}(-b)$.

2. For $i \neq j$, show that

$$I = D_i(-1)T_{ji}(1)T_{ij}(-1)T_{ji}(1)P_{ij}$$
$$= D_j(-1)T_{ij}(1)T_{ji}(-1)T_{ij}(1)P_{ij}.$$

   Hence an elementary matrix of type I is a product of elementary matrices of type II and of type III.

3. Compute

$$\det \begin{pmatrix} 2 & 0 & 0 & 1 & -2 \\ 0 & 1 & 3 & -3 & 1 \\ -2 & -3 & -5 & 2 & -1 \\ 4 & -4 & 4 & -6 & 0 \\ 2 & -2 & 1 & 0 & 1 \end{pmatrix}.$$

   Please use whatever method we have discussed about the determinant so far to do this problem. Do not use "cofactor expansion" to solve the determinant. The discussion of "cofactor expansion" is yet to come.

3.3 Reduced row echelon form and applications

The concept of reduced row echelon form (or even non-reduced row echelon form) is helpful for solving systems of linear equations, finding inverses and much more. In this section we will review how to find a reduced row echelon form for a matrix over a field, and discuss how the process fares for a matrix over a ring.

Row echelon form and reduced row echelon form

Remember that a matrix is in **row echelon form** if it satisfies the following requirements.

- Any nonzero rows are above the zero rows.

- The farthest left nonzero entry of each nonzero row is 1 and is called the **pivot** of that row. The pivot is strictly to the right of the pivot in the previous row.

The second point actually says that the entries below the pivots must be zero!

A matrix is in **reduced row echelon form** if it satisfies one additional condition:

- The pivot in each nonzero row is the only nonzero entry in its column.

The point here is that the entries above the pivots must be zero as well.

Before we discuss the main result in this section, we first make a rather obvious observation on a matrix in general.

Lemma 3.3.1. *If a square matrix over a ring has a zero row (or column), it is not invertible.*

Proof. Let A be a square matrix whose i-th row is zero. If A is invertible, there is another matrix B such that $AB = I$, the identity matrix. However, from definition of the product of matrices, we can see that the i-th row of AB must also be zero, a contradiction. A similar argument works for the column case. $\square$

Proposition 3.3.2. *Let $A \in M_{m \times n}(F)$ where F is a field. We can find a series of elementary matrices $E_1, E_2, \ldots, E_N$ such that*

$$(3.3.1) \qquad B = E_N \cdots E_2 E_1 A$$

is in reduced row echelon form. The matrix B is called a reduced row-echelon form of A. We also have that

(3.3.2) $$A = E_1^{-1} E_2^{-1} \cdots E_N^{-1} B.$$

Furthermore, suppose A is a square matrix of size n. The following conditions are equivalent.

(i) *The matrix A is invertible.*

(ii) *The reduced row echelon form B has no zero rows.*

(iii) *The reduced row echelon form $B = I_n$.*

(iv) *The determinant of A is nonzero.*

If A is invertible, then $A^{-1} = E_N \cdots E_2 E_1$ and $A = E_1^{-1} E_2^{-1} \cdots E_N^{-1}$. Hence, an square matrix over a field is invertible if and only if it is a product of elementary matrices.

Proof. For the first part, it suffices to show that after a series of elementary row operations, we may change A into a reduced row echelon form. We will do this by induction on m.

If A is the zero matrix, it already is in reduced row echelon form. We will assume that $A \neq 0$.

If $m = 1$, A is of the form

$$\begin{pmatrix} 0 & \cdots & 0 & a & * & \cdots & * \end{pmatrix}$$

where $a \neq 0$. If we multiply this row by a^{-1}, it will be in reduce row echelon form. We now may assume that $m > 1$.

Find the farthest left column containing a nonzero entry. Move the row containing the said nonzero entry to the first row. Suppose the said nonzero entry is u. Multiply the first row by u^{-1}. Now this row is lead by 1, which is a pivot. If there is any nonzero entry a under this pivot in the i-th row, we may add $(-a)$ times the first row to the i-th row to cancel that entry. In this way we may clear away all nonzero entries under the pivot. Now the matrix is changed into the form

$$\begin{pmatrix} 0\cdots0 & 1 & *\cdots* \\ & 0 & \\ 0 & \vdots & C \\ & 0 & \end{pmatrix}$$

The matrix C has one less row than A. By induction we may perform a series of elementary row operations to turn C into a reduced row echelon form. These elementary row operations will not affect the first row and the zero columns to the left of C. The entries below and above the pivots in the reduced row echelon form of C are all 0. We may perform elementary row operations using these pivots of C to clear away the nonzero entries in the first row in the bigger matrix. We have thus reached a reduced row echelon form B for A. For each elementary row operation performed we may find a corresponding elementary matrix. We may describe the process of transforming A into a reduced row echelon form B by an identity as in (3.3.1). From (3.3.1) we may also obtain the identity (3.3.2).

For the second part, let A be a square matrix of size n.

We first show that "(ii) $\Leftrightarrow$ (iii)". Remember that B is in reduced row echelon form. The number of pivots in B is the same as the number of nonzero rows in B. When B has a zero row, clearly $B \neq I_n$. If B has no zero rows, B has n pivots. There must be a pivot in each of the n columns of B. In each column the entries other than the pivot are all zero. Thus $B = I_n$.

"(i) $\Rightarrow$ (ii)": By (3.3.1) we have that B is invertible. This says that B cannot have any zero rows from Lemma 3.3.1.

"(iii) $\Rightarrow$ (i)": Suppose $B = I_n$. From (3.3.2) we have that

$$(3.3.3) \qquad A = E_1^{-1} E_2^{-1} \cdots E_N^{-1}$$

is a product of invertible matrices by Proposition 3.2.3. This shows that A is invertible by Exercise 4, §2.3. In fact,

$$A^{-1} = (E_1^{-1} E_2^{-1} \cdots E_N^{-1})^{-1} = E_N \cdots E_2 E_1.$$

At this point we have shown that (i) $\Leftrightarrow$ (ii) $\Leftrightarrow$ (iii). It remains to show that "(i) $\Leftrightarrow$ (iv)".

"(i) $\Rightarrow$ (iv)": If A is invertible, then

$$\begin{aligned} \det A &= \det(E_1^{-1} E_2^{-1} \cdots E_N^{-1}) \\ &= \det E_1^{-1} \det E_2^{-1} \cdots \det E_N^{-1} \neq 0 \end{aligned}$$

from (3.3.3), Proposition 3.2.3, Corollary 3.2.5, and Corollary 3.2.6.

"(iv) $\Rightarrow$ (i)": Suppose A is not invertible. Then B has a zero row by (ii). Thus $\det B = 0$ by Proposition 3.1.13(b). We have that

$$\begin{aligned} \det A &= \det(E_1^{-1} E_2^{-1} \cdots E_N^{-1} B) \\ &= \det E_1^{-1} \det E_2^{-1} \cdots \det E_N^{-1} \det B = 0 \end{aligned}$$

from (3.3.2), Proposition 3.2.3, and Corollary 3.2.5. The rest of the proposition follows immediately.

When A is invertible, A is a product of elementary matrices by (3.3.3). Conversely, A is invertible by Exercise 4, §2.3.   □

If we change all the "reduced row echelon form" in Proposition 3.3.2 to "row echelon form", the first part of the proposition still holds true. A square matrix over a field is invertible if and only if (any) one of its row echelon form contains no zero rows. We leave the details to the reader.

We now summarize the process of finding a row echelon form of a matrix over a field. For this we only need elementary row operations. First, locate the candidate for the first pivot. Move it to the first row and modify the first row so that the pivot is 1. Use it to clear away the entries below the pivot. Locate the candidate for the second pivot. Process it so that this pivot is 1 and lies in the second row. Use it to clear away the nonzero entries below the pivot if we only want to find the row echelon form. If we want to find the reduced row echelon form, then use the pivot to clear away the nonzero entries above the pivot as well. Repeat the process until there are no longer nonzero rows to deal with.

In actual practice, one may improvise to facilitate the process of reaching a (reduced) row echelon form when one becomes experienced in such a task.

Example 3.3.3. Find a reduced row echelon form for the matrix

$$A = \begin{pmatrix} 0 & 1 & 1 & 1 \\ 2 & 4 & 1 & 2 \\ 3 & 3 & 1 & 1 \end{pmatrix} \in M_{3 \times 4}(\mathbb{Z}_5).$$

Solution. The $(2,1)$-entry is a "potential" pivot. But instead of multiplying the second row by 2^{-1}, we will use a different approach for the first step. We have

$$A \rightsquigarrow \begin{pmatrix} 0 & 1 & 1 & 1 \\ 2 & 4 & 1 & 2 \\ 1 & 4 & 0 & 4 \end{pmatrix} \quad \text{Note that } -1 = 4 \text{ in } \mathbb{Z}_5. \\ \rightarrow \text{row 3} - \text{row 2}$$

$$\rightsquigarrow \begin{pmatrix} \boxed{1} & 4 & 0 & 4 \\ 2 & 4 & 1 & 2 \\ 0 & 1 & 1 & 1 \end{pmatrix}, \text{ switching row 1 and row 3}$$

$$\rightsquigarrow \begin{pmatrix} \boxed{1} & 4 & 0 & 4 \\ 0 & \boxed{1} & 1 & 4 \\ 0 & 1 & 1 & 1 \end{pmatrix} \quad \rightarrow \text{row 2} - 2(\text{row 1})$$

$$\rightsquigarrow \begin{pmatrix} \boxed{1} & 4 & 0 & 4 \\ 0 & \boxed{1} & 1 & 4 \\ 0 & 0 & 0 & 2 \end{pmatrix} \quad \rightarrow \text{row 3} - \text{row 2}$$

$$\rightsquigarrow \begin{pmatrix} \boxed{1} & 4 & 0 & 4 \\ 0 & \boxed{1} & 1 & 4 \\ 0 & 0 & 0 & \boxed{1} \end{pmatrix} \quad \rightarrow 2^{-1}(\text{row 3}).$$

We have reached a row echelon form for A. The pivots lie in the first, the second and the fourth columns. So far, we only use the pivots to clear away the entries below the pivots. To reach the reduced row echelon form, we proceed to clear away the entries above the pivots. We have

$$A \rightsquigarrow \begin{pmatrix} \boxed{1} & 0 & 1 & 3 \\ 0 & \boxed{1} & 1 & 4 \\ 0 & 0 & 0 & \boxed{1} \end{pmatrix} \quad \rightarrow \text{row 1} - 4(\text{row 2})$$

$$\rightsquigarrow \begin{pmatrix} \boxed{1} & 0 & 1 & 0 \\ 0 & \boxed{1} & 1 & 0 \\ 0 & 0 & 0 & \boxed{1} \end{pmatrix} \quad \begin{array}{l} \rightarrow \text{row 1} - 3(\text{row 3}) \\ \rightarrow \text{row 2} - 4(\text{row 3}) \end{array}$$

This is a reduced row echelon form for A. $\diamond$

The discussion above only works if we consider matrices over a field. The next example shows how the process of finding reduced row echelon form fails in general.

Example 3.3.4. Suppose it is possible to apply elementary row operations to the matrix

$$A = \begin{pmatrix} 1 & 2 \\ 3 & 4 \end{pmatrix} \in M_2(\mathbb{Z})$$

so that it reaches a reduced row echelon form B. Then there exist elementary matrices E_i, $i = 1, 2, \ldots, n$, such that

$$B = E_N \cdots E_1 A.$$

The matrix B must be one of the following types:

$$\begin{pmatrix} 0 & 0 \\ 0 & 0 \end{pmatrix}, \quad \begin{pmatrix} 0 & 1 \\ 0 & 0 \end{pmatrix}, \quad \begin{pmatrix} 1 & * \\ 0 & 0 \end{pmatrix} \quad \text{or} \quad \begin{pmatrix} 1 & 0 \\ 0 & 1 \end{pmatrix},$$

whose determinant is either 0 or 1. By Lemma 3.2.4, the determinants of elementary matrices are ± 1 in $\mathbb{Z}$. From Corollary 3.2.5 we have

$$\det B = \det E_n \cdots \det E_1 \det A = \pm \det A.$$

The determinant of A is either 0 or ± 1, which is clearly not the case for A.

Solving a system of linear equations

We may use reduced echelon form to help us solve a system of linear equations in the following way.

Suppose given a general system of linear equations over a field F:

$$(3.3.4) \quad \begin{cases} a_{11}x_1 + a_{12}x_2 + \cdots + a_{1n}x_n = b_1 \\ a_{21}x_1 + a_{22}x_2 + \ldots + a_{2n}x_n = b_2 \\ \cdots\cdots\cdots\cdots\cdots\cdots\cdots\cdots\cdots\cdots\cdots\cdots\cdots\cdots\cdots \\ \cdots\cdots\cdots\cdots\cdots\cdots\cdots\cdots\cdots\cdots\cdots\cdots\cdots\cdots \\ a_{m1}x_1 + a_{m2}x_2 + \ldots + a_{mn}x_n = b_m \end{cases}$$

This system may be expressed in matrix form as $A\mathbf{x} = \mathbf{b}$ where

$$A = \begin{pmatrix} a_{11} & a_{12} & \cdots & a_{1n} \\ a_{21} & a_{22} & \cdots & a_{2n} \\ \vdots & \vdots & \ddots & \vdots \\ a_{m1} & a_{m2} & \cdots & a_{mn} \end{pmatrix}, \quad \mathbf{x} = \begin{pmatrix} x_1 \\ x_2 \\ \vdots \\ x_n \end{pmatrix} \quad \text{and} \quad \mathbf{b} = \begin{pmatrix} b_1 \\ b_2 \\ \vdots \\ b_n \end{pmatrix}.$$

The matrix A is called the **coefficient matrix** of the system, $\mathbf{x}$ the variable vector and $\mathbf{b}$ the constant variable. To solve the system in (3.3.4), we may set up the **augmented matrix** $\left(\; A \mid \mathbf{b} \; \right)$ for the system. We construct the augmented matrix by adding the constant vector $\mathbf{b}$ to the right of A with a dividing line in between for our convenience. It is clear that we may also retrieve the system of linear equations from a given augmented matrix.

When we perform an elementary row operation to the augmented matrix, we are doing a corresponding action to the system of linear equations. For example, when we exchange two rows in the augmented matrix, it is equivalent to exchanging the two corresponding linear equations. By doing so, the solution space to the system is unchanged. We leave it to the reader to verify that we do not change the solution space to the system no matter which type of elementary row operations we perform on the augmented matrix.

We may perform various elementary row operations to the augmented matrix until A reaches a reduced row echelon form. This process is called **Gaussian elimination.**

Suppose the augmented matrix is transformed into the form $\left(\; B \mid \mathbf{c} \; \right)$ where B is a reduced row echelon form of A. In case there is a nonzero entry c in $\mathbf{c}$ at a zero row of B, we say this linear system is **inconsistent**. The solution space to an inconsistent system of linear equations is *empty* for the augmented matrix $\left(\; B \mid \mathbf{c} \; \right)$ gives a contradictory equation $0 = c$. On the other hand, when no such thing happens, we say the linear system is **consistent**. The solution space to a consistent system is transparent as we shall see in the following example.

Example 3.3.5. Solve the system of linear equations

$$\begin{cases} & 2y & + & 4z & = & 1 \\ 2x & + & 4y & + & 2z & = & 1 \\ 3x & + & 3y & + & z & = & 1 \end{cases}$$

over $\mathbb{Q}$.

Solution. The given system may be expressed in matrix form

$$(3.3.5) \qquad A \begin{pmatrix} x \\ y \\ z \end{pmatrix} = \begin{pmatrix} 1 \\ 1 \\ 1 \end{pmatrix}, \qquad \text{where } A = \begin{pmatrix} 0 & 2 & 4 \\ 2 & 4 & 2 \\ 3 & 3 & 1 \end{pmatrix}.$$

We may perform elementary row operations to the augmented matrix for the given system until the matrix A on the left is in reduced row echelon form. We have that

$$
\left(\begin{array}{ccc|c}
0 & 2 & 4 & 1 \\
2 & 4 & 2 & 1 \\
3 & 3 & 1 & 1
\end{array}\right)
\rightsquigarrow
\left(\begin{array}{ccc|c}
0 & 1 & 2 & 1/2 \\
1 & 2 & 1 & 1/2 \\
3 & 3 & 1 & 1
\end{array}\right)
\rightsquigarrow
\left(\begin{array}{ccc|c}
1 & 2 & 1 & 1/2 \\
0 & 1 & 2 & 1/2 \\
3 & 3 & 1 & 1
\end{array}\right)
$$

$$
\rightsquigarrow
\left(\begin{array}{ccc|c}
1 & 2 & 1 & 1/2 \\
0 & 1 & 2 & 1/2 \\
0 & -3 & -2 & -1/2
\end{array}\right)
\rightsquigarrow
\left(\begin{array}{ccc|c}
1 & 0 & -3 & -1/2 \\
0 & 1 & 2 & 1/2 \\
0 & 0 & 4 & 1
\end{array}\right)
$$

$$
\rightsquigarrow
\left(\begin{array}{ccc|c}
1 & 0 & -3 & -1/2 \\
0 & 1 & 2 & 1/2 \\
0 & 0 & 1 & 1/4
\end{array}\right)
\rightsquigarrow
\left(\begin{array}{ccc|c}
1 & 0 & 0 & 1/4 \\
0 & 1 & 0 & 0 \\
0 & 0 & 1 & 1/4
\end{array}\right).
$$

This gives the solution

$$
\begin{pmatrix} x \\ y \\ z \end{pmatrix} = \begin{pmatrix} 1/4 \\ 0 \\ 1/4 \end{pmatrix}.
$$

$\diamond$

How to find the inverse of a square matrix

Proposition 3.3.2 suggests a method to compute the inverse of an invertible matrix. We may apply various elementary row operations to the augmented matrix

$$
\left(A \mid I_n\right)
$$

until the matrix A in the left hand box becomes I_n. Find elementary matrices $E_1, E_2, \ldots, E_N$ such that $E_N \cdots E_2 E_1 A = I_n$ as in (3.3.1). Then

$$
E_N \cdots E_2 E_1 \left(A \mid I_n\right) = \left(E_N \cdots E_2 E_1 A \mid E_N \cdots E_2 E_1 I_n\right)
$$
$$
= \left(I_n \mid E_N \cdots E_2 E_1\right).
$$

By Proposition 3.3.2, the inverse of A is $E_N \cdots E_2 E_1$, which sits in the right hand box of the resulting matrix.

Example 3.3.6. Solve the system of linear equations

$$
\begin{cases}
& 2y + 4z = 1 \\
2x + 4y + 2z = 1 \\
3x + 3y + z = 1
\end{cases}.
$$

Solution. This is a revisit to Example 3.3.5. We will reuse the notation there.

If the coefficient matrix A is invertible, then

$$\begin{pmatrix} x \\ y \\ z \end{pmatrix} = A^{-1} \begin{pmatrix} 1 \\ 1 \\ 1 \end{pmatrix}$$

from (3.3.5). We simply need to find A^{-1}. We have

$$\left(\begin{array}{ccc|ccc} 0 & 2 & 4 & 1 & 0 & 0 \\ 2 & 4 & 2 & 0 & 1 & 0 \\ 3 & 3 & 1 & 0 & 0 & 1 \end{array} \right) \rightsquigarrow \left(\begin{array}{ccc|ccc} 0 & 1 & 2 & 1/2 & 0 & 0 \\ 1 & 2 & 1 & 0 & 1/2 & 0 \\ 3 & 3 & 1 & 0 & 0 & 1 \end{array} \right)$$

$$\rightsquigarrow \left(\begin{array}{ccc|ccc} 1 & 2 & 1 & 0 & 1/2 & 0 \\ 0 & 1 & 2 & 1/2 & 0 & 0 \\ 3 & 3 & 1 & 0 & 0 & 1 \end{array} \right) \rightsquigarrow \left(\begin{array}{ccc|ccc} 1 & 2 & 1 & 0 & 1/2 & 0 \\ 0 & 1 & 2 & 1/2 & 0 & 0 \\ 0 & -3 & -2 & 0 & -3/2 & 1 \end{array} \right)$$

$$\rightsquigarrow \left(\begin{array}{ccc|ccc} 1 & 0 & -3 & -1 & 1/2 & 0 \\ 0 & 1 & 2 & 1/2 & 0 & 0 \\ 0 & 0 & 4 & 3/2 & -3/2 & 1 \end{array} \right) \rightsquigarrow \left(\begin{array}{ccc|ccc} 1 & 0 & -3 & -1 & 1/2 & 0 \\ 0 & 1 & 2 & 1/2 & 0 & 0 \\ 0 & 0 & 1 & 3/8 & -3/8 & 1/4 \end{array} \right)$$

$$\rightsquigarrow \left(\begin{array}{ccc|ccc} 1 & 0 & 0 & 1/8 & -5/8 & 3/4 \\ 0 & 1 & 0 & -1/4 & 3/4 & -1/2 \\ 0 & 0 & 1 & 3/8 & -3/8 & 1/4 \end{array} \right).$$

This gives the solution

$$\begin{pmatrix} x_1 \\ x_2 \\ x_3 \end{pmatrix} = A^{-1} \begin{pmatrix} 1 \\ 1 \\ 1 \end{pmatrix} = \begin{pmatrix} 1/8 & -5/8 & 3/4 \\ -1/4 & 3/4 & -1/2 \\ 3/8 & -3/8 & 1/4 \end{pmatrix} \begin{pmatrix} 1 \\ 1 \\ 1 \end{pmatrix} = \begin{pmatrix} 1/4 \\ 0 \\ 1/4 \end{pmatrix}.$$

$\diamond$

The determinant of the product of two matrices

The following result is very useful.

Proposition 3.3.7. *Let A and B be square matrices of the same size over a field F. Then $\det AB = \det A \det B$.*

Proof. Find elementary matrices $E_1, \ldots, E_N$ such that $A = E_N \cdots E_1 C$ where C is a reduced row echelon form by Proposition 3.3.2.

Case 1. Suppose A is not invertible. From Proposition 3.3.2, we have $\det A = 0$ and C is of the form

$$\begin{pmatrix} & * & \\ 0 & \cdots & 0 \end{pmatrix},$$

which has a zero row at the bottom. This implies that

$$AB = E_N \cdots E_1 \begin{pmatrix} & * & \\ 0 & \cdots & 0 \end{pmatrix} B = E_N \cdots E_1 \begin{pmatrix} & * & \\ 0 & \cdots & 0 \end{pmatrix}.$$

Hence

$$\det AB = \det E_N \cdots \det E_1 \cdot 0 = 0 = \det A \det B$$

by Proposition 3.1.13(b), Corollary 3.2.5 and the fact that $\det A = 0$.

Case 2. Suppose A is invertible. By Proposition 3.3.2, C is the identity matrix and $A = E_N \cdots E_1$. Hence

$$\begin{aligned} \det AB &= \det(E_N \cdots E_1 B) = \det E_N \cdots \det E_1 \cdot \det B \\ &= \det(E_N \cdots E_1) \det B = \det A \det B \end{aligned}$$

by Corollary 3.2.5. $\qquad\qquad\square$

Now we are ready to prove the general case.

Theorem 3.3.8. *Let A and B be square matrices of the same size over a ring R. Then $\det AB = \det A \det B$.*

Proof. Let $A = (a_{ij})$ and $B = (b_{ij})$ be two matrices of size n with entries in R. Let $\{x_{ij}, y_{ij} : i, j = 1, \ldots, n\}$ be $2n^2$ indeterminates over $\mathbb{Z}$ and let $S = \mathbb{Z}[x_{ij}, y_{ij} : i, j = 1, \ldots, n]$. Let $\phi \colon S \to R$ be the ring homomorphism sending x_{ij} to a_{ij} and y_{ij} to b_{ij}. Let $X = (x_{ij})$ and $Y = (y_{ij})$. Consider X and Y as matrices with entries in Q, the quotient field of S. Then we have that $\det XY = \det X \det Y$ in Q by Proposition 3.3.7. However, $\det XY$, $\det X$ and $\det Y$ are in S which is a subring of Q. We also have $\det XY = \det X \det Y$ in S. Thus

$$\begin{aligned} \det AB &= \phi(\det XY) = \phi(\det X \det Y) \\ &= \phi(\det X) \phi(\det Y) = \det A \det B \end{aligned}$$

by Lemma 3.1.12. $\qquad\qquad\square$

Exercises 3.3

For the following exercises F denotes a field.

1. State the definition for column echelon form and reduced column echelon form. Restate and prove the column echelon form version of Proposition 3.3.2.

2. Find the reduced row echelon form of

$$\begin{pmatrix} 0 & 1 & 1 & 1 & 1 \\ 1 & 0 & 1 & 1 & 1 \\ 1 & 1 & 0 & 1 & 1 \\ 1 & 1 & 1 & 0 & 1 \end{pmatrix}$$

over $\mathbb{Z}_2$.

3. Let $\mathbf{v}_1, \mathbf{v}_2, \ldots, \mathbf{v}_m$ be row vectors in F^n and let $A = \begin{pmatrix} \mathbf{v}_1 \\ \vdots \\ \mathbf{v}_m \end{pmatrix}$. Let B be a reduced echelon form of A.

   (a) Show that the row space of A is the same as the row space of B.

   (b) Show that the nonzero rows in B are linearly independent over F, and hence form a basis for the row space of A. Conclude that the row rank of A is the number of nonzero rows in B.

   (c) Show that the column rank of A is the same as the column rank of B. (Hint: View the elementary matrices as matrices representing isomorphisms.)

   (d) Show that the column rank of B is the number of nonzero rows in B.

   (e) Conclude that $\operatorname{col} \operatorname{rk} A = \operatorname{row} \operatorname{rk} A$. Hence we simply call the column (or row) rank of A the **rank** of A, denoted $\operatorname{rk} A$. (*Cf.* Proposition 2.3.12.)

4. Explain why the rank of a matrix is unaffected by elementary operations.

5. Find the rank of

$$\begin{pmatrix} 1 & 2 & 3 & 4 \\ 2 & 3 & 4 & 5 \\ 3 & 4 & 5 & 6 \end{pmatrix}$$

over $\mathbb{R}$.

6. Does the set

$$\left\{ \begin{pmatrix} 2 \\ 0 \\ -2 \\ 4 \end{pmatrix}, \begin{pmatrix} 0 \\ 1 \\ -3 \\ -4 \end{pmatrix}, \begin{pmatrix} 0 \\ 3 \\ -5 \\ 4 \end{pmatrix}, \begin{pmatrix} 1 \\ -3 \\ 2 \\ -6 \end{pmatrix} \right\}$$

form a basis for $\mathbb{R}^4$?

7. Is there a 3×5 matrix A over $\mathbb{R}$ whose first, third and fourth columns of A are

$$\begin{pmatrix} 1 \\ -1 \\ 3 \end{pmatrix}, \begin{pmatrix} 0 \\ -1 \\ 1 \end{pmatrix} \text{ and } \begin{pmatrix} 1 \\ -2 \\ 0 \end{pmatrix}$$

respectively, and whose reduced echelon form is

$$\begin{pmatrix} 1 & 2 & 0 & 0 & 2 \\ 0 & 0 & 1 & 0 & -3 \\ 0 & 0 & 0 & 1 & 1 \end{pmatrix}.$$

8. Is there a 3×5 matrix A over $\mathbb{R}$ whose first, third and fourth columns of A are

$$\begin{pmatrix} 1 \\ -1 \\ 3 \end{pmatrix}, \begin{pmatrix} 0 \\ -1 \\ 1 \end{pmatrix} \text{ and } \begin{pmatrix} 3 \\ 2 \\ 4 \end{pmatrix}$$

respectively, and whose reduced echelon form is

$$\begin{pmatrix} 1 & 2 & 0 & 0 & 2 \\ 0 & 0 & 1 & 0 & -3 \\ 0 & 0 & 0 & 1 & 1 \end{pmatrix}.$$

9. Show that the reduced row echelon form of a matrix over a field is unique.

10. Let $A \in M_{m \times n}(F)$ and $\operatorname{rk} A = r$. Show that A is equivalent to

$$\begin{pmatrix} I_r & \\ & \mathbf{0}_{(m-r) \times (n-r)} \end{pmatrix}.$$

3.4 Determinant and invertible matrices

In §3.3 most of the results regarding matrices with entries in a field hangs on to the fact that we can perform elementary row operations on a matrix to reach a reduced row echelon form. In Example 3.3.4 we saw that this is not the case in general. Thus for the general case, we need to develop a different approach.

Throughout this section R denotes a ring.

Minors and cofactors

Definition 3.4.1. Let $A = \left(a_{ij}\right)_{n \times n}$ be a square matrix of size n over R. For $1 \leq i, j \leq n$, we define the (i, j)-**minor** of A, denoted M_{ij}, to be the $(n-1) \times (n-1)$ matrix obtained by deleting the i-th row and the j-th column from A. The (i, j)-**cofactor** of A, denoted A_{ij}, is defined to be

$$A_{ij} = (-1)^{i+j} \det M_{ij}.$$

Lemma 3.4.2. *Let* $A \in M_n(R)$ *and let* $a \in R$. *Let*

$$B = \begin{pmatrix} A & \begin{matrix} * \\ \vdots \\ * \end{matrix} \\ 0 \ \cdots \ 0 & a \end{pmatrix}.$$

Then $\det B = a \det A$.

Proof. Let $A = (a_{ij})_{n \times n}$ and $B = (b_{ij})_{(n+1) \times (n+1)}$. Then

$$b_{ij} = \begin{cases} a_{ij}, & \text{for } 1 \leq i,\, j \leq n; \\ 0, & \text{for } i = n+1 \text{ and } 1 \leq j \leq n; \\ 0, & \text{for } j = n+1 \text{ and } 1 \leq i \leq n; \\ a, & \text{for } i = j = n+1. \end{cases}$$

We have that

$$
\begin{aligned}
\det B &= \sum_{\sigma \in S_{n+1}} (\operatorname{sg} \sigma)\, b_{1\sigma(1)} b_{2\sigma(2)} \cdots b_{n\sigma(n)} b_{n+1,\sigma(n+1)} \\
&= \sum_{\substack{\sigma \in S_{n+1} \\ \sigma(n+1)=n+1}} (\operatorname{sg} \sigma)\, a_{1\sigma(1)} a_{2\sigma(2)} \cdots a_{n\sigma(n)} a \\
&= a \sum_{\sigma \in S_n} (\operatorname{sg} \sigma)\, a_{1\sigma(1)} a_{2\sigma(2)} \cdots a_{n\sigma(n)} \\
&= a \det A.
\end{aligned}
$$

The argument holds because $b_{n+1,j} = 0$ for $j \neq n+1$. $\qquad\square$

Proposition 3.4.3 (Cofactor expansion). *Let*

$$
A = \left(a_{ij}\right)_{n \times n} \in M_n(R)
$$

for $n \geq 2$. For $i = 1, 2, \ldots, n$, we have

$$
(3.4.1) \qquad \det A = a_{i1} A_{i1} + a_{i2} A_{i2} + \cdots + a_{in} A_{in}.
$$

*This is called **cofactor expansion** of $\det A$ along the i-th row of A. For $j = 1, 2, \ldots, n$, we have*

$$
(3.4.2) \qquad \det A = a_{1j} A_{1j} + a_{2j} A_{2j} + \cdots + a_{nj} A_{nj}.
$$

This is called the cofactor expansion of $\det A$ along the j-th column of A.

Proof. From Proposition 3.1.13(f) we have that $\det A = \sum_j \det B_j$ where B_j is the matrix obtained by replacing the i-th row of A with the row $(0, \ldots, 0, a_{ij}, 0, \ldots, 0)$. To find $\det B_j$ for each j, we first switch the i-th row of B_j with the $(i+1)$-th row. Then switch the $(i+1)$-th row with the $(i+2)$-th row. Repeat this process until the original i-th row reaches the bottom row. This is done by applying $n-i$ elementary row operations of type I. Now switch the j-th column with the $(j+1)$-th column. Then switch the $(j+1)$-th column with the $(j+2)$-th column. Repeat this process until the original j-th column reaches the last column. This is done by applying

$n - j$ elementary column operations of type I. We have that

$$\det B_j = (-1)^{n-i+n-j} \det \begin{pmatrix} M_{ij} & & & * \\ & & & \vdots \\ & & & * \\ 0 & \cdots & 0 & a_{ij} \end{pmatrix}$$

$$= (-1)^{i+j} a_{ij} \det M_{ij} = a_{ij} A_{ij}$$

by Lemma 3.4.2. We thus have the formula in (3.4.1).

The proof for the formula in (3.4.2) is similar.   □

Corollary 3.4.4. *Let* $A = \left(a_{ij}\right)_{n \times n} \in M_n(R)$ *where* $n \geq 2$.
Let $1 \leq i, j, k \leq n$. *If* $i \neq k$, *then*

$$a_{i1}A_{k1} + a_{i2}A_{k2} + \cdots + a_{in}A_{kn} = 0.$$

If $j \neq k$, *then*

$$a_{1j}A_{1k} + a_{2j}A_{2k} + \cdots + a_{nj}A_{nk} = 0.$$

Proof. If $i \neq k$, $a_{i1}A_{k1} + a_{i2}A_{k2} + \cdots + a_{in}A_{kn}$ is the determinant of the matrix obtained by replacing the k-th row of A with its i-th row. This matrix has repeated rows. Its determinant is 0 by Proposition 3.1.13(e).

The proof of the column version is similar.   □

Example 3.4.5. Use cofactor expansion to compute $\det A$ where

$$A = \begin{pmatrix} 2 & 0 & 0 & 1 \\ 0 & 1 & 3 & -3 \\ -2 & -3 & -5 & 2 \\ 4 & -4 & 4 & -6 \end{pmatrix} \in M_4(\mathbb{Z}).$$

Solution. We choose the first row of A to do the cofactor expansion. From Proposition 3.4.3 we have that $\det A = 2A_{11} + A_{14}$. Since

$$A_{11} = \det M_{11} = \det \begin{pmatrix} 1 & 3 & -3 \\ -3 & -5 & 2 \\ -4 & 4 & -6 \end{pmatrix} = 40$$

and

$$A_{14} = -\det M_{14} = -\det \begin{pmatrix} 0 & 1 & 3 \\ -2 & -3 & -5 \\ 4 & -4 & 4 \end{pmatrix} = -48,$$

we have $\det A = 2(40) - 48 = 32$.   ◇

Definition 3.4.6. Let $A \in M_n(R)$. We define the **adjoint** of the matrix A, denoted adj A, to be $(A_{ij})^{\text{tr}} = (A_{ji})$.

Example 3.4.7. Compute adj A where

$$A = \begin{pmatrix} 1 & 1 & 1 \\ 1 & 2 & 2 \\ 1 & 2 & 3 \end{pmatrix} \in M_3(\mathbb{Z}).$$

Solution. First we compute the A_{ij}'s:

$$A_{11} = \det \begin{pmatrix} 2 & 2 \\ 2 & 3 \end{pmatrix} = 2; \quad A_{12} = -\det \begin{pmatrix} 1 & 2 \\ 1 & 3 \end{pmatrix} = -1;$$

$$A_{13} = \det \begin{pmatrix} 1 & 2 \\ 1 & 2 \end{pmatrix} = 0; \quad A_{21} = -\det \begin{pmatrix} 1 & 1 \\ 2 & 3 \end{pmatrix} = -1;$$

$$A_{22} = \det \begin{pmatrix} 1 & 1 \\ 1 & 3 \end{pmatrix} = 2; \quad A_{23} = -\det \begin{pmatrix} 1 & 1 \\ 1 & 2 \end{pmatrix} = -1;$$

$$A_{31} = \det \begin{pmatrix} 1 & 1 \\ 2 & 2 \end{pmatrix} = 0; \quad A_{32} = -\det \begin{pmatrix} 1 & 1 \\ 1 & 2 \end{pmatrix} = -1;$$

$$A_{33} = \det \begin{pmatrix} 1 & 1 \\ 1 & 2 \end{pmatrix} = 1.$$

Hence

$$\text{adj } A = \begin{pmatrix} 2 & -1 & 0 \\ -1 & 2 & -1 \\ 0 & -1 & 1 \end{pmatrix}^{\text{tr}} = \begin{pmatrix} 2 & -1 & 0 \\ -1 & 2 & -1 \\ 0 & -1 & 1 \end{pmatrix}.$$

Here the adjoint of A happens to be a symmetric matrix, a coincidence. ◇

Combining Proposition 3.4.3 and Corollary 3.4.4, we have

$$a_{i1}A_{k1} + a_{i2}A_{k2} + \cdots + a_{in}A_{kn} = \begin{cases} \det A, & \text{if } i = k, \\ 0, & \text{if } i \neq k; \end{cases}$$

$$a_{1j}A_{1k} + a_{2j}A_{2k} + \cdots + a_{nj}A_{nk} = \begin{cases} \det A, & \text{if } j = k, \\ 0, & \text{if } j \neq k \end{cases}$$

for $i, j, k = 1, \ldots, n$. This gives us the following result.

Corollary 3.4.8. *Let $A \in M_n(R)$. Then*

$$A(\operatorname{adj} A) = (\operatorname{adj} A)A = (\det A)I_n = \begin{pmatrix} \det A & & & \\ & \det A & & \mathbf{0} \\ & & \ddots & \\ \mathbf{0} & & & \det A \end{pmatrix}.$$

Invertible matrices

Now we can make the following conclusion on invertible matrices.

Theorem 3.4.9. *Let A and B be square matrices of size n with entries in a ring R.*

(a) *The matrix A is invertible in $M_n(R)$ if and only if $\det A$ is a unit in R. In this case $\det A^{-1} = (\det A)^{-1}$.*

(b) *Furthermore, $BA = I_n$ if and only if $AB = I_n$. In either case*

$$B = \frac{1}{\det A} \operatorname{adj} A$$

is the inverse of A.

Proof. (a) The "only if" part: Suppose A is invertible. From Theorem 3.3.8 we have $\det A \det A^{-1} = \det(AA^{-1}) = \det I_n = 1$ in R. This shows that $\det A$ is a unit in R and $\det A^{-1} = (\det A)^{-1}$.

The "if" part: Suppose $\det A$ is a unit in R. From Corollary 3.4.8 we have

$$A\big[(\det A)^{-1}\operatorname{adj} A\big] = (\det A)^{-1}\big[A\operatorname{adj} A\big] = (\det A)^{-1}\big[(\det A)I_n\big] = I_n;$$
$$\big[(\det A)^{-1}\operatorname{adj} A\big]A = (\det A)^{-1}\big[(\operatorname{adj} A)A\big] = (\det A)^{-1}\big[(\det A)I_n\big] = I_n.$$

Hence $(\det A)^{-1}\operatorname{adj} A$ is the inverse of A and A is invertible over R.

(b) The "if" part: Assume $AB = I_n$. Then $\det A \det B = 1$ from Theorem 3.3.8. Hence $\det A$ is a unit in R. This implies that

$$B = IB = \left[(\det A)^{-1} \operatorname{adj} A\right] AB = (\det A)^{-1} \operatorname{adj} A.$$

Thus $BA = \left[(\det A)^{-1} \operatorname{adj} A\right] A = (\det A)^{-1} \det A I_n = I_n$ as well.

The "only if" part follows by symmetry. $\square$

Theorem 3.4.9 seems to give us a good formula for finding the inverse of an invertible matrix. However, to find the adjoint of a matrix is often not practical. Theorem 3.4.9 is more of a theorem for theoretical purposes. However, it is possible to apply elementary row operations to transform an invertible matrix over $\mathbb{Z}$ (or the Gaussian integer ring $\mathbb{Z}[i]$, or $F[x]$ where x is a variable over the field F) into the identity matrix. Please see Exercise 5, §4.2 for further details. Assuming this fact, we will use the method of finding the inverse of an invertible matrix in §3.3 for matrices with entries in rings such as $\mathbb{Z}$, $\mathbb{Z}[i]$ or $F[x]$.

Example 3.4.10. Determine whether the following matrices are invertible, and find the inverses for the ones that are indeed invertible.

(a) $A = \begin{pmatrix} 1 & 2 & 3 \\ 2 & 1 & 3 \\ 3 & 1 & 2 \end{pmatrix} \in M_3(\mathbb{Z})$;

(b) $B = \begin{pmatrix} 1+i & 2i+2 & 2-i \\ i & 1 & 0 \\ 0 & 1 & -i \end{pmatrix} \in M_3(\mathbb{Z}[i])$;

(c) $C = \begin{pmatrix} x+3 & 0 & x+1 & 0 \\ x & 1 & 0 & x \\ x+1 & 2x-3 & 1 & 2x^2-3x \\ x+2 & 0 & x^2+1 & 1 \end{pmatrix} \in M_4(\mathbb{Q}[x])$ where x is a variable over $\mathbb{Q}$.

Solution. (a) Since $\det A = 6$ is not a unit in $\mathbb{Z}$, the matrix A is not invertible over $\mathbb{Z}$.

(b) We use the cofactor expansion along the third row to find $\det B$:

$$\det B = -\det \begin{pmatrix} 1+i & 2-i \\ i & 0 \end{pmatrix} - i \det \begin{pmatrix} 1+i & 2i+2 \\ i & 1 \end{pmatrix} = -i$$

is a unit in $\mathbb{Z}[i]$. This shows that B is invertible over $\mathbb{Z}[i]$. We proceed to find the inverse of B:

$$\begin{pmatrix} 1+i & 2i+2 & 2-i & 1 & 0 & 0 \\ i & 1 & 0 & 0 & 1 & 0 \\ 0 & 1 & -i & 0 & 0 & 1 \end{pmatrix}$$

$$\rightsquigarrow \begin{pmatrix} 1 & -i & 0 & 0 & -i & 0 \\ 1+i & 2i+2 & 2-i & 1 & 0 & 0 \\ 0 & 1 & -i & 0 & 0 & 1 \end{pmatrix}, \quad \begin{array}{l} \rightarrow (-i)(\text{row } 2) \\ \text{switching row 1 and row 2} \end{array}$$

$$\rightsquigarrow \begin{pmatrix} 1 & -i & 0 & 0 & -i & 0 \\ 0 & 1+3i & 2-i & 1 & -1+i & 0 \\ 0 & 1 & -i & 0 & 0 & 1 \end{pmatrix} \quad \rightarrow \text{row } 2 - (1+i)(\text{row } 1)$$

$$\rightsquigarrow \begin{pmatrix} 1 & -i & 0 & 0 & -i & 0 \\ 0 & 1 & -i & 0 & 0 & 1 \\ 0 & 1+3i & 2-i & 1 & -1+i & 0 \end{pmatrix}, \quad \text{switching row 2 and row 3}$$

$$\rightsquigarrow \begin{pmatrix} 1 & 0 & 1 & 0 & -i & i \\ 0 & 1 & -i & 0 & 0 & 1 \\ 0 & 0 & -1 & 1 & -1+i & -1-3i \end{pmatrix} \quad \begin{array}{l} \rightarrow \text{row } 1 + i(\text{row } 2) \\ \rightarrow \text{row } 3 - (1+3i)(\text{row } 2) \end{array}$$

$$\rightsquigarrow \begin{pmatrix} 1 & 0 & 1 & 0 & -i & i \\ 0 & 1 & -i & 0 & 0 & 1 \\ 0 & 0 & 1 & -1 & 1-i & 1+3i \end{pmatrix} \quad \rightarrow -(\text{row } 3)$$

$$\rightsquigarrow \begin{pmatrix} 1 & 0 & 0 & 1 & -1 & -1-2i \\ 0 & 1 & 0 & -i & 1+i & -2+i \\ 0 & 0 & 1 & -1 & 1-i & 1+3i \end{pmatrix} \quad \begin{array}{l} \rightarrow \text{row } 1 - \text{row } 3 \\ \rightarrow \text{row } 2 + i(\text{row } 3) \end{array}$$

Just as expected, the method for finding the inverse of a matrix in §3.3 follows through. Hence

$$B^{-1} = \begin{pmatrix} 1 & -1 & -1-2i \\ -i & 1+i & -2+i \\ -1 & 1-i & 1+3i \end{pmatrix}$$

(c) We use the cofactor expansion along the first row to find

$$\det C = (x+3)\det \begin{pmatrix} 1 & 0 & x \\ 2x-3 & 1 & 2x^2-3x \\ 0 & x^2+1 & 1 \end{pmatrix}$$

$$+ (x+1)\det \begin{pmatrix} x & 1 & x \\ x+1 & 2x-3 & 2x^2-3x \\ x+2 & 0 & 1 \end{pmatrix}$$

$$= x + 3 + (x+1)(2x^2 - 4x - 1) = 2x^3 - 2x^2 - 6x + 2.$$

The determinant of C is not a unit in $\mathbb{Q}[x]$. Hence C is not invertible. ⋄

Proposition 3.4.11. *Let R be a commutative ring (this result is not true if R is not noncommutative). If $R^r \cong R^s$, then $r = s$.*

Proof. Suppose the claim is not true. Without loss of generality we may assume $r < s$.

Suppose given an isomorphism $\varphi \colon R^s \to R^r$. Let $A_{r \times s}$ be the matrix of φ with respect to the standard bases of R^s and R^r. Let $B_{s \times r}$ be the matrix of φ^{-1} with respect to the standard bases of R^r and R^s. From Proposition 2.3.6 we have that $AB = I_r$ and $BA = I_s$. Add $s - r$ zero rows below A and $s - r$ zero columns on the right side of B to form two new matrices

$$A' = \begin{pmatrix} A_{r \times s} \\ \mathbf{0}_{(s-r) \times s} \end{pmatrix}_{s \times s} \qquad \text{and} \qquad B' = \begin{pmatrix} B_{s \times r} & \mathbf{0}_{s \times (s-r)} \end{pmatrix}_{s \times s}.$$

Note that $B'A' = (BA) = I_s$. From Theorem 3.4.9(b), we should have $A'B' = I_s$ as well. However,

$$A'B' = \begin{pmatrix} A_{r \times s} \\ \mathbf{0}_{(s-r) \times s} \end{pmatrix} \begin{pmatrix} B_{s \times r} & \mathbf{0}_{s \times (s-r)} \end{pmatrix} = \begin{pmatrix} I_r & \mathbf{0}_{r \times (s-r)} \\ \mathbf{0}_{(s-r) \times r} & \mathbf{0}_{(s-r) \times (s-r)} \end{pmatrix},$$

a contradiction. □

This proposition tells us that if M is a free R-module with a finite basis, we may define the **rank** of M, denoted $\mathrm{rk}_R M$ or simply $\mathrm{rk}\, M$ when R is understood, to be the size of any finite free basis for M.

Exercises 3.4

Throughout this set of exercises, R denotes a ring.

1. Suppose A and B are two square matrices over R such that $AB = 0$. Is it true that $\det A = 0$ or $\det B = 0$?

2. Let $T\colon \mathbb{R}^5 \to \mathbb{R}^5$ be the linear transformation over $\mathbb{R}$ such that

$$T(e_1) = e_2 - 2e_3 + e_5;$$
$$T(e_2) = e_1 - e_2 + e_4 - e_5;$$
$$T(e_3) = e_1 - 2e_2 + e_3 + e_4 - e_5;$$
$$T(e_4) = -e_1 + 3e_5;$$
$$T(e_5) = 2e_1 + e_2 + 2e_3 - e_4 - 2e_5.$$

Can you use the result in this section to determine whether T is an isomorphism? If instead, T is a $\mathbb{Z}$-linear map from $\mathbb{Z}^5$ to $\mathbb{Z}^5$, is T an isomorphism?

3. Let A and B be square matrices (not necessarily of the same size) and C be a matrix of appropriate size over R. Show that

$$\det \begin{pmatrix} A & C \\ & B \end{pmatrix} = \det A \det B.$$

4. Let A, B, C and D be square matrices of the same size such that C and D commute and D is invertible. Show that

$$\det \begin{pmatrix} A & B \\ C & D \end{pmatrix} = \det(AD - BC).$$

(Hint: Multiply on the right by $\begin{pmatrix} D & 0 \\ -C & D^{-1} \end{pmatrix}$.) Is this assertion still true if D is not invertible or if C and D do not commute?

5. Let $A \in M_n(R)$ and let λ be an indeterminate over R. Define the **characteristic polynomial** of A to be $\det(\lambda I - A)$. Compute the characteristic polynomial of the following matrices.

(a) $A = \begin{pmatrix} r & 1 & & & \\ & r & 1 & & \\ & & \ddots & \ddots & \\ & & & r & 1 \\ & & & & r \end{pmatrix}$

$$\text{(b)} \quad A = \begin{pmatrix} 0 & 1 & 0 & \cdots & 0 \\ 0 & 0 & 1 & \cdots & 0 \\ \vdots & \vdots & \vdots & \ddots & \vdots \\ 0 & 0 & 0 & \cdots & 1 \\ \alpha_{n-1} & \alpha_{n-2} & \alpha_{n-3} & \cdots & \alpha_0 \end{pmatrix}.$$

6. Let x_{ij}, $i, j = 1, \ldots, n$, be n^2 indeterminates over $\mathbb{Z}$ and let

$$R = \mathbb{Z}[\, x_{ij} : 1 \leq i, \; j \leq n\,].$$

Let $A = (x_{ij}) \in M_n(R)$. Is $\det A$ nonzero? Is A invertible?

3.5 Rank

In the exercises of §3.3 we described how to compute the rank of a matrix with entries in a field. It all depends on the fact that we can apply elementary row operations to a matrix until it is in row echelon form. The rank of a matrix is thus the number of nonzero rows in a row echelon form of the said matrix. However, in general it is not always possible to apply elementary row operations to transform a matrix into row echelon form. In this section we are going to give a more general characterization on the rank of a matrix.

Throughout this section R denotes a ring.

Minor ideals

Let A be an $m \times n$ matrix with entries in R. When we delete $(m - i)$ rows and $(n - i)$ columns from A, we obtain an $i \times i$ submatrix of A. The determinant of this submatrix is called an i-**minor** of A. (Note that this "minor" is different from the "minor" in Definition 3.4.1.) We will use $I_i(A)$ to denote the ideal generated by all the i-minors of A over R, and this is called a **minor ideal** of A. In particular, $I_1(A)$ is the ideal generated by the entries of A.

Example 3.5.1. Let

$$A = \begin{pmatrix} 12 & -15 & 30 \\ 6 & 9 & -6 \end{pmatrix}.$$

If A is a matrix over $\mathbb{Q}$, the column space of A is $\mathbb{Q}^2$ since the first two column vectors are linearly independent over $\mathbb{Q}$ and thus form a basis for $\mathbb{Q}^2$. The third column vector of A can be generated by the first two over $\mathbb{Q}$. However, if A is a matrix over $\mathbb{Z}$, the situation is totally different. The first two column vectors are still linearly independent over $\mathbb{Z}$, but they do not generate $\mathbb{Z}^2$. In particular, the third column vector is not generated by the first two column vectors. In fact, no column of A can be generated by the other two columns over $\mathbb{Z}$. (See Exercise 2.)

The 1-minors of A are the entries of A. Over $\mathbb{Z}$,

$$I_1(A) = (12,\ -15,\ 30,\ 6,\ 9,\ -6) = (3).$$

The 2-minors of A are

$$\begin{cases} \det \begin{pmatrix} 12 & -15 \\ 6 & 9 \end{pmatrix} = 6 \cdot 3 \cdot 11, \\[2mm] \det \begin{pmatrix} 12 & 30 \\ 6 & -6 \end{pmatrix} = -6 \cdot 6 \cdot 7, \\[2mm] \det \begin{pmatrix} -15 & 30 \\ 9 & -6 \end{pmatrix} = -6 \cdot 3 \cdot 10. \end{cases}$$

Hence $I_2(A) = (18)$.

Note that if we consider A as a matrix over $\mathbb{Q}$, then $I_1(A) = I_2(A) = (1)$.

Example 3.5.2. Let $R = \mathbb{Z}[x, y, z, w, u, v]$ be the polynomial ring of 6 indeterminates over $\mathbb{Z}$ and let

$$A = \begin{pmatrix} x & z & u \\ y & w & v \end{pmatrix}.$$

Then $I_1(A) = (x, y, z, w, u, v)$ and $I_2(A) = (xw - yz,\ xv - yu,\ zv - wu)$.

Let Q be the quotient field of R. If instead we view A as a matrix over Q, then $I_1(A) = I_2(A) = (1)$.

Let A be an $m \times n$ matrix over R. It is clear in general that

$$I_1(A) \supseteq I_2(A) \supseteq I_3(A) \supseteq \cdots$$

since the i-minors are all R-linear combinations of the $(i-1)$-minors using cofactor expansion. For this reason, we define $I_i(A) = R$ for $i \leq 0$ and

$I_i(A) = (0)$ for $i > \min\{m, n\}$ as a convention. For $1 \leq i \leq \min\{m, n\}$, note that $I_i(A) = (0)$ if and only if all the i-minors of A are 0.

Let A be a matrix with entries in a ring. We will use $A^{(i_1, i_2, \ldots, i_k)}$ to denote the matrix obtained by only retaining row i_1, ..., row i_k of A. We will also use $A_{(j_1, j_2, \ldots, j_\ell)}$ to denote the matrix obtained by only retaining column j_1, ..., column j_ℓ of A. We will use $A^{(i_1, i_2, \ldots, i_k)}_{(j_1, j_2, \ldots, j_\ell)}$ to denote $\left[A^{(i_1, i_2, \ldots, i_k)} \right]_{(j_1, j_2, \ldots, j_\ell)} = \left[A_{(j_1, j_2, \ldots, j_\ell)} \right]^{(i_1, i_2, \ldots, i_k)}$.

Lemma 3.5.3. *Let $A_{\ell \times m}$ and $B_{m \times n}$ be matrices with entries in the ring R. Then*

$$(AB)^{(i_1, i_2, \ldots, i_s)}_{(j_1, j_2, \ldots, j_t)} = A^{(i_1, i_2, \ldots, i_s)} B_{(j_1, j_2, \ldots, j_t)}$$

for $1 \leq i_1 < \cdots < i_s \leq \ell$ and $1 \leq j_1 < \cdots < j_t \leq n$.

Proof. It follows form the fact that the (i, j)-entry of AB are derived from the i-th row of A and the j-th column of B. □

We use an example to help demonstrate Lemma 3.5.3.

Example 3.5.4. Let

$$A = \begin{pmatrix} 1 & 2 & 3 \\ 1 & 1 & 1 \\ 3 & 2 & 1 \end{pmatrix} \quad \text{and} \quad B = \begin{pmatrix} 1 & 1 & 1 & 1 \\ 1 & 2 & 2 & 2 \\ 1 & 2 & 3 & 4 \end{pmatrix}$$

be matrices over $\mathbb{Z}$. Then

$$AB = \begin{pmatrix} 6 & 11 & 14 & 17 \\ 3 & 5 & 6 & 7 \\ 6 & 9 & 10 & 11 \end{pmatrix}.$$

Note that

$$A^{(2,3)} B_{(1,3,4)} = \begin{pmatrix} 1 & 1 & 1 \\ 3 & 2 & 1 \end{pmatrix} \begin{pmatrix} 1 & 1 & 1 \\ 1 & 2 & 2 \\ 1 & 3 & 4 \end{pmatrix} = \begin{pmatrix} 3 & 6 & 7 \\ 6 & 10 & 11 \end{pmatrix} = (AB)^{(2,3)}_{(1,3,4)}.$$

Proposition 3.5.5. *Let $A \in M_{i \times m}(R)$ and $B \in M_{m \times i}(R)$. If $i \leq m$, then*

$$\det(AB) = \sum_{1 \leq j_1 < \cdots < j_i \leq m} \left(\det A_{(j_1, j_2, \ldots, j_i)} \right) \left(\det B^{(j_1, j_2, \ldots, j_i)} \right).$$

If $i > m$ then $\det AB = 0$.

Note that Theorem 3.3.8 is the special case of Proposition 3.5.5 when $i = m$.

Example 3.5.6. Before proving Proposition 3.5.5, we will verify that it is true for
$$A = \begin{pmatrix} 1 & 2 & 3 \\ 3 & 3 & 3 \end{pmatrix} \quad \text{and} \quad B = \begin{pmatrix} 1 & 1 \\ 1 & 2 \\ 1 & 3 \end{pmatrix}.$$

We have
$$\det \begin{pmatrix} 1 & 2 & 3 \\ 3 & 3 & 3 \end{pmatrix} \begin{pmatrix} 1 & 1 \\ 1 & 2 \\ 1 & 3 \end{pmatrix} = \det \begin{pmatrix} 6 & 14 \\ 9 & 18 \end{pmatrix} = 3 \cdot 2 \det \begin{pmatrix} 2 & 7 \\ 3 & 9 \end{pmatrix} = -18.$$

According to Proposition 3.5.5, this determinant should be the sum of the following terms:

$$(j_1, j_2) = (1,2): \quad \det \begin{pmatrix} 1 & 2 \\ 3 & 3 \end{pmatrix} \det \begin{pmatrix} 1 & 1 \\ 1 & 2 \end{pmatrix} = (-3)1 = -3;$$

$$(j_1, j_2) = (1,3): \quad \det \begin{pmatrix} 1 & 3 \\ 3 & 3 \end{pmatrix} \det \begin{pmatrix} 1 & 1 \\ 1 & 3 \end{pmatrix} = (-6)2 = -12;$$

$$(j_1, j_2) = (2,3): \quad \det \begin{pmatrix} 2 & 3 \\ 3 & 3 \end{pmatrix} \det \begin{pmatrix} 1 & 2 \\ 1 & 3 \end{pmatrix} = (-3)1 = -3.$$

Since $-18 = -3 - 12 - 3$, we can see that indeed Proposition 3.5.5 holds in this example.

Proof of Proposition 3.5.5. Suppose $i > m$. Let

$$A' = \begin{pmatrix} A_{i \times m} & \mathbf{0}_{i \times (i-m)} \end{pmatrix}_{i \times i} \quad \text{and} \quad B' = \begin{pmatrix} B_{m \times i} \\ \mathbf{0}_{(i-m) \times i} \end{pmatrix}_{i \times i}.$$

It follows that $A'B' = AB$ from Exercise 15, §2.3. Hence

$$\det AB = \det A'B' = \det A' \det B' = 0$$

by Theorem 3.3.8 and Proposition 3.1.13(b).

We may now assume that $i \leq m$. We will prove Proposition 3.5.5 by induction on i.

The case $i = 1$ follows from the fact that $\det AB = \sum_{j=1}^{m} a_{1j} b_{j1}$ while $\det A_{(j)} = a_{1j}$ and $\det B^{(j)} = b_{j1}$.

Assume $i > 1$. Let $AB = C = (c_{k\ell})$. Using the cofactor expansion along the i-th row of C and Lemma 3.5.3, we have

$$\det C = \sum_{\ell=1}^{i} (-1)^{i+\ell} c_{i\ell} \det M_{i\ell}$$

where $M_{i\ell} = A^{(1,\dots,i-1)} B_{(1,\dots,\widehat{\ell},\dots,i)}$ is the (i,ℓ)-minor of $C = AB$. Let $\mathbf{J} = \{j_1, j_2, \dots, j_{i-1}\}$ be a subset of $\{1, 2, \dots, m\}$ with $i-1$ elements. We will further assume $1 \le j_1 < j_2 < \cdots < j_{i-1} \le m$. Let S be the set consisting of all such $\mathbf{J}$'s. Since the minor matrices $M_{i\ell}$ have only $(i-1)$ rows, by induction we have

$$\det C = \sum_{\ell=1}^{i} (-1)^{i+\ell} \left[\left(\sum_{k=1}^{m} a_{ik} b_{k\ell} \right) \det M_{i\ell} \right]$$

$$= \sum_{\ell=1}^{i} \sum_{k=1}^{m} (-1)^{i+\ell} a_{ik} b_{k\ell} \left(\sum_{\mathbf{J} \in S} \det A^{(1,\dots,i-1)}_{(j_1, j_2, \dots, j_{i-1})} \det B^{(j_1, j_2, \dots, j_{i-1})}_{(1,\dots,\widehat{\ell},\dots,i)} \right)$$

$$= \sum_{\ell=1}^{i} \sum_{k=1}^{m} \sum_{\mathbf{J} \in S} (-1)^{i+\ell} a_{ik} \det A^{(1,\dots,i-1)}_{(j_1, j_2, \dots, j_{i-1})} b_{k\ell} \det B^{(j_1, j_2, \dots, j_{i-1})}_{(1,\dots,\widehat{\ell},\dots,i)}$$

$$= \sum_{k=1}^{m} \sum_{\mathbf{J} \in S} \sum_{\ell=1}^{i} (-1)^{i+\ell} a_{ik} \det A^{(1,\dots,i-1)}_{(j_1, j_2, \dots, j_{i-1})} b_{k\ell} \det B^{(j_1, j_2, \dots, j_{i-1})}_{(1,\dots,\widehat{\ell},\dots,i)}$$

$$= \sum_{k=1}^{m} \sum_{\mathbf{J} \in S} \left[(-1)^{i} a_{ik} \det A^{(1,\dots,i-1)}_{(j_1, j_2, \dots, j_{i-1})} \sum_{\ell=1}^{i} (-1)^{\ell} b_{k\ell} \det B^{(j_1, j_2, \dots, j_{i-1})}_{(1,\dots,\widehat{\ell},\dots,i)} \right]$$

$$= \sum_{k=1}^{m} \sum_{\mathbf{J} \in S} \left[(-1)^{i+\mathbf{J}_k+1} a_{ik} \det A^{(1,\dots,i-1)}_{(j_1, j_2, \dots, j_{i-1})} \right.$$

$$\left. \times \left(\sum_{\ell=1}^{i} (-1)^{\mathbf{J}_k+1+\ell} b_{k\ell} \det B^{(j_1, j_2, \dots, j_{i-1})}_{(1,\dots,\widehat{\ell},\dots,i)} \right) \right]$$

where $\mathbf{J}_k = \#\{j \in \mathbf{J} : j < k\}$. Note that

$$\sum_{\ell=1}^{i} (-1)^{\mathbf{J}_k+1+\ell} b_{k\ell} \det B^{(j_1, j_2, \dots, j_{i-1})}_{(1,\dots,\widehat{\ell},\dots,i)}$$

is the determinant of $B^{(j_1,\dots,k,\dots,j_{i-1})}$ if $j_t < k < j_{t+1}$ for some t. If k equals one of the j_t's, this term is the determinant of some matrix with two identical rows and is thus zero. Thus,

$$\det C = \sum_{k=1}^{m} \left(\sum_{\substack{\mathbf{J} \in S \\ k \notin \mathbf{J}}} (-1)^{i+\mathbf{J}_k+1} a_{ik} \det A^{(1,\ldots,i-1)}_{(j_1, j_2, \ldots, j_{i-1})} \det B^{(j_1,\ldots,k,\ldots,j_{i-1})} \right).$$

In the summand, $\{j_1, \ldots, k, \ldots, j_{i-1}\}$ is a set of i elements. Let $\mathfrak{J}$ be an i-element subset $\{j_1, j_2, \ldots, j_i\}$ of $\{1, \ldots, m\}$. We will further assume that $1 \leq j_1 < j_2 < \cdots < j_i \leq m$. Let T be the set of all such $\mathfrak{J}$'s. We may now rearrange the sum which is $\det C$ as follows:

$$\det C = \sum_{\mathfrak{J} \in T} \det B^{(j_1, j_2, \ldots, j_i)} \left(\sum_{t=1}^{i} (-1)^{i+t} a_{ij_t} \det A^{(1,\ldots,i-1)}_{(j_1,\ldots,\widehat{j_t},\ldots,j_i)} \right)$$

$$= \sum_{\mathfrak{J} \in S} \det B^{(j_1, j_2, \ldots, j_i)} \det A_{(j_1, j_2, \ldots, j_i)}.$$

Hence the result. $\qquad\qquad\qquad\qquad\qquad\qquad\qquad\qquad\qquad\qquad\qquad\quad\square$

Proposition 3.5.7. *Let $A \in M_{\ell \times m}(R)$ and $B \in M_{m \times n}(R)$. Then*

$$I_i(AB) \subseteq I_i(A) \qquad and \qquad I_i(AB) \subseteq I_i(B)$$

for all i.

Proof. It suffices to check the case when $1 \leq i \leq \min\{m, n\}$.

From Lemma 3.5.3, an i-minor of AB is the determinant of the product of two matrices, the first being a submatrix of i-rows of A and the second being a submatrix of i columns of B. From Proposition 3.5.5, we know this determinant is a sum in which each summand is a product of an i-minor of A and an i-minor of B. Thus each i-minor of AB is in both $I_i(A)$ and $I_i(B)$. Thus the ideal $I_i(AB)$ is contained in both $I_i(A)$ and in $I_i(B)$. $\quad\square$

Corollary 3.5.8. *Let $A \in M_{m \times n}(R)$. If U and V are invertible matrices of size m and n respectively, then $I_i(UAV) = I_i(A)$ for all i.*

Proof. From Proposition 3.5.7, we have that $I_i(UAV) \subseteq I_i(A)$. Similarly we also have $I_i(A) = I_i(U^{-1}(UAV)V^{-1}) \subseteq I_i(UAV)$. Hence the result. $\quad\square$

Corollary 3.5.9. *Let $A \in M_{m \times n}(F)$ where F is a field. If $\mathrm{rk}\, A = r$, then*

$$I_i(A) = \begin{cases} F, & \text{if } i \leq r; \\ (0), & \text{otherwise.} \end{cases}$$

Proof. Using Exercise 10, §3.3, we have that A is equivalent to

$$
B = \begin{pmatrix}
1 & & & \\
& \ddots & & \\
& & 1 & \\
& & & \mathbf{0}_{(m-r) \times (n-r)}
\end{pmatrix}
$$

with only r nonzero rows. It is clear that $I_i(B) = (1)$ for $i \leq r$ and $I_i(B) = (0)$ otherwise. Since $I_i(A) = I_i(B)$ for all i by Corollary 3.5.8, we have the result. □

Rank

Corollary 3.5.9 is the basis of the following definition.

Definition 3.5.10. Let A be a matrix over R. Define the **rank** of A over R to be

$$
\mathrm{rk}_R A = \max\{\, i \in \mathbb{Z} : I_i(A) \neq 0 \,\}.
$$

We may simply write $\mathrm{rk}\, A$ for $\mathrm{rk}_R A$ when R is understood.

Corollary 3.5.9 tells us that the rank of a matrix with entries in a field in the sense defined in §3.3 coincides with the rank defined in Definition 3.5.10.

Note that $I_0(A) = (1)$ by default for any matrix A. Thus the rank of any matrix is at least 0 by definition. However, only the zero matrices (of any size) would be of rank 0, since I_1 is generated by the entries of the matrix in question. By default, $\mathrm{rk}\, A \leq \min\{m, n\}$ if A is an $m \times n$ matrix.

When the rank of a matrix is r, it means that one of the r-minors of A is nonzero while all the $(r+1)$-minors of A are zero. Thus we have the following result.

Corollary 3.5.11. *Let S be a ring and let R be a subring of S. If A is a matrix with entries in R, then $\mathrm{rk}_R A = \mathrm{rk}_S A$.*

Proof. The result follows because an i-minor of A is 0 in R if and only if it is 0 in S. □

Example 3.5.12. Let's compute the rank of A over $\mathbb{Z}$ where

$$A = \begin{pmatrix} 0 & 1 & 2 \\ 1 & 2 & 3 \\ 2 & 3 & 4 \\ 3 & 4 & 5 \end{pmatrix}.$$

From Corollary 3.5.11, the rank of A over $\mathbb{Z}$ equals the rank of A over $\mathbb{Q}$. Hence we may use the reduced row echelon form of A over $\mathbb{Q}$ to find the rank of A over $\mathbb{Z}$. Or instead, note that the four 3-minors of A

$$\det \begin{pmatrix} 1 & 2 & 3 \\ 2 & 3 & 4 \\ 3 & 4 & 5 \end{pmatrix}, \quad \det \begin{pmatrix} 0 & 1 & 2 \\ 2 & 3 & 4 \\ 3 & 4 & 5 \end{pmatrix}, \quad \det \begin{pmatrix} 0 & 1 & 2 \\ 1 & 2 & 3 \\ 3 & 4 & 5 \end{pmatrix}, \quad \det \begin{pmatrix} 0 & 1 & 2 \\ 1 & 2 & 3 \\ 2 & 3 & 4 \end{pmatrix}$$

are all zero while the top left 2-minor of A is

$$\det \begin{pmatrix} 0 & 1 \\ 1 & 2 \end{pmatrix} = -1 \neq 0.$$

We conclude that $\mathrm{rk}_{\mathbb{Z}} A = \mathrm{rk}_{\mathbb{Q}} A = 2$.

Example 3.5.13. In this example we first compute the rank of A over $\mathbb{Z}_{12}$ where

$$A = \begin{pmatrix} 3 & 6 & 0 \\ 3 & 2 & 4 \end{pmatrix}.$$

Since A is nonzero, its rank is at least 1. Observe that

$$\begin{cases} \det \begin{pmatrix} 3 & 6 \\ 3 & 2 \end{pmatrix} = -12 = 0, \\[2mm] \det \begin{pmatrix} 3 & 0 \\ 3 & 4 \end{pmatrix} = 12 = 0, \\[2mm] \det \begin{pmatrix} 6 & 0 \\ 2 & 4 \end{pmatrix} = 24 = 0 \end{cases}$$

in $\mathbb{Z}_{12}$. We conclude that $\mathrm{rk}_{\mathbb{Z}_{12}} A = 1$. Note that $\mathrm{rk}_{\mathbb{Z}} A = 2$ if we consider A as a matrix over $\mathbb{Z}$.

If a matrix is of rank 1 over a field, each of its columns would have been a multiple of one of the columns. This claim is certainly not true for the matrix A over a non-domain such as $\mathbb{Z}_{12}$. This claim will not be true in general even when A is a matrix over a domain. See Exercise 4.

Exercises 3.5

1. Let A be an $\ell \times m$ matrix, B be an $m \times n$ matrix, both with entries in a ring R. Show that rk $AB \leq \min\{\ell,\ m,\ n\}$.

2. Let A be the matrix in Example 3.5.1. Show that no column of A can be generated by the other two columns in A over $\mathbb{Z}$.

3. Let A and Q be as in Example 3.5.2. Then $\mathrm{rk}_Q A = 2$. Express $\begin{pmatrix} u \\ v \end{pmatrix}$ as a linear combination of $\begin{pmatrix} x \\ y \end{pmatrix}$ and $\begin{pmatrix} z \\ w \end{pmatrix}$ over Q.

4. Give an example of a 2×2 matrix A over $\mathbb{Z}$ such that $\mathrm{rk}_{\mathbb{Z}} A = 1$ and no column of A is an integer multiple of the other column.

5. Let $R = \mathbb{Z}[x, y, z, w, u, v]$ be the polynomial ring of 6 variables over $\mathbb{Z}$ and let
$$A = \begin{pmatrix} \overline{x} & \overline{z} & \overline{u} \\ \overline{y} & \overline{w} & \overline{v} \end{pmatrix}$$
be a matrix over $S = R/(xw - yz,\ xv - yu,\ zv - wu)$. Find $\mathrm{rk}_S A$.

6. Compute the rank r of A where
$$A = \begin{pmatrix} -w & & & & & x \\ & -w & & & & y \\ & & -w & z & & \\ -z & & & x & & \\ & -z & & y & & \\ -y & & x & & & \end{pmatrix}$$
is a matrix over $\mathbb{Z}[x, y, z, w]$, the polynomial rings of 4 variables over $\mathbb{Z}$. Find I_r as well.

7. Let x_{ij}, $1 \leq i \leq m$ and $1 \leq j \leq n$, be indeterminates over $\mathbb{Z}$ and let

$$R = \mathbb{Z}[\, x_{ij} : 1 \leq i \leq m, \ 1 \leq j \leq n\,].$$

Compute $\mathrm{rk}_R \, (x_{ij})_{m \times n}$.

For the following exercises, we assume all matrices are with entries in a field F.

8. Let A and B be two square matrices of size n and $AB = 0$. Show that $\mathrm{rk}\, A + \mathrm{rk}\, B \leq n$.

9. For any square matrix A of size n, show that there is a square matrix B such that $AB = 0$ and $\mathrm{rk}\, A + \mathrm{rk}\, B = n$.

10. Let A be a square matrix of rank 1.

   (a) Show that there is a unique $\alpha \in F$ such that $A^2 = \alpha A$. (Hint: The matrix A is a product of a column vector and a row vector.)

   (b) Show that $1 - A$ is invertible if $\alpha \neq 1$. (Hint: Show that the kernel of $1 - A$ is trivial.)

CHAPTER 4

Canonical Forms

In this chapter we will study *linear endomorphisms* on vector spaces. A linear endomorphism is a linear map from a vector space to itself. The domain is the codomain! Thus it is reasonable to use the same ordered basis for both the domain and the codomain. We will discuss how to find an ordered basis so that the matrix representing a given linear endomorphism is as simple as possible. This "simplest" matrix will be called a canonical form of the linear endomorphism. The main result of this chapter is that we may use *the* canonical form (there are two types, the *rational* form and/or the *Jordan* form) to classify a linear endomorphism. However, the most logical and the most natural way to analyze a linear endomorphism on a vector space V is, surprisingly, to view V as a module over the polynomial ring of one variable over the given field. This gives us a great incentive to discuss *the fundamental theorem of finitely generated modules over a PID*, a major result in module theory. Not only the result of this theorem is important, the process of proving this theorem is a great example on how to analyze a module.

4.1   Equivalent matrices

Throughout this section R denotes a ring.

Base change matrices and invertible matrices

First we review some results from §2.3. Suppose given an R-linear map $f\colon R^n \to R^m$. Let β and γ be ordered bases of R^n and R^m respectively. We have a matrix A representing f with respect to β and γ. Suppose given another two ordered bases β' and γ' for R^n and R^m respectively. Let P be the base change matrix from β' to β and Q be the base change matrix from γ' to γ. The matrix representing f with respect to β' and γ' is $Q^{-1}AP$. We are simply restating Corollary 2.3.10 here.

It is easy to recognize a base change matrix. Corollary 2.3.8 says that a base change matrix is always invertible. The converse is also true according to Exercise 5 in §2.3. But for clarity of discussion, we will show the converse in the following proposition.

Proposition 4.1.1. *Let $P = (p_{ij})_{n \times n}$ be an invertible matrix in $M_n(R)$. Suppose $\beta = (u_1, u_2, \ldots, u_n)$ is an ordered basis of R^n. Let*

$$v_j = p_{1j}u_1 + p_{2j}u_2 + \cdots + p_{nj}u_n, \qquad for\ j = 1, 2, \ldots, n.$$

Then $\beta' = (v_1, v_2, \ldots, v_n)$ is also an ordered basis of R^n. Hence the invertible matrix P can be viewed as a base change matrix from β' to β.

Proof. Since P is invertible, we may express P^{-1} as the matrix $(q_{ij})_{n \times n}$ with entries in R.

For each i with $1 \le i \le n$, we have

$$\sum_{k=1}^n q_{ki}v_k = \sum_{k=1}^n \left(q_{ki} \sum_{j=1}^n p_{jk}u_j \right) = \sum_{j=1}^n \left(\sum_{k=1}^n p_{jk}q_{ki} \right) u_j.$$

Here $a_i = \sum_{k=1}^n p_{jk}q_{ki}$ is the (j, i)-entry of the matrix $PP^{-1} = I_n$. Thus $a_i = 1$ if $j = i$ and $a_i = 0$ if $j \ne i$. Hence

$$u_i = \sum_{k=1}^n q_{ki}v_k, \qquad for\ i = 1, 2, \ldots, n.$$

We have that β' generates a basis of R^n over R. Thus β' generates R^n over R. Next we will proceed to show that β' is linearly independent over R. We cannot use Proposition 1.4.9 for this purpose because it only works for finite dimensional vector spaces over a *field*.

Now suppose $\sum_{j=1}^n c_j v_j = 0$ where $c_j \in R$ for all j. Then

$$0 = \sum_{j=1}^n c_j \left(\sum_{i=1}^n p_{ij} u_i \right) = \sum_{i=1}^n \left(\sum_{j=1}^n p_{ij} c_j \right) u_i.$$

This implies that $\sum_{j=1}^n p_{ij} c_j = 0$ for all i by the linear independence of β. In matrix form, we have

$$P \begin{pmatrix} c_1 \\ \vdots \\ c_n \end{pmatrix} = \begin{pmatrix} 0 \\ \vdots \\ 0 \end{pmatrix} \implies \begin{pmatrix} c_1 \\ \vdots \\ c_n \end{pmatrix} = P^{-1} P \begin{pmatrix} c_1 \\ \vdots \\ c_n \end{pmatrix} = P^{-1} \begin{pmatrix} 0 \\ \vdots \\ 0 \end{pmatrix} = \begin{pmatrix} 0 \\ \vdots \\ 0 \end{pmatrix}.$$

This shows that $v_1, v_2, \ldots, v_n$ are linearly independent over R. We conclude that β' is a basis for R^n over R. It also follows that P is the base change matrix from β' to β. $\square$

Example 4.1.2. Let $T: \mathbb{R}^2 \to \mathbb{R}^3$ be the $\mathbb{R}$-linear transformation sending (x, y) to $(2x + 3y, x - y, 5x - 2y)$.

(a) Find the matrix representing T with respect to the standard bases $\beta = (e_1, e_2)$ and $\gamma = (e_1, e_2, e_3)$.

(b) Let $\beta' = ((1, 0), (1, 1))$ and $\gamma' = ((0, 1, 1), (1, 0, 1), (1, 1, 0))$. Show that β' and γ' are ordered bases of $\mathbb{R}^2$ and $\mathbb{R}^3$ respectively.

(c) Find the matrix representing T with respect to β' and γ'.

Solution. (a) Since $T(e_1) = (2, 1, 5)$ and $T(e_2) = (3, -1, -2)$, the matrix representing T with respect to the respective standard bases is

$$A = \begin{pmatrix} 2 & 3 \\ 1 & -1 \\ 5 & -2 \end{pmatrix}.$$

(b) The determinant of the matrix

$$P = \begin{pmatrix} 1 & 1 \\ 0 & 1 \end{pmatrix}$$

is 1. Hence P is invertible by Theorem 3.4.9. Proposition 4.1.1 tells us that β' is also a basis for $\mathbb{R}^2$ over $\mathbb{R}$. Similarly, since

$$\det Q = \det \begin{pmatrix} 0 & 1 & 1 \\ 1 & 0 & 1 \\ 1 & 1 & 0 \end{pmatrix} = 2$$

is invertible in $\mathbb{R}$, we know that γ' is a basis for $\mathbb{R}^3$ over $\mathbb{R}$.

(c) The matrix P is the base change matrix from β' to β. The matrix Q is the base change matrix from γ' to γ. The matrix representing T with respect to β' and γ' is

$$Q^{-1}AP = \frac{1}{2} \begin{pmatrix} -1 & 1 & 1 \\ 1 & -1 & 1 \\ 1 & 1 & -1 \end{pmatrix} \begin{pmatrix} 2 & 3 \\ 1 & -1 \\ 5 & -2 \end{pmatrix} \begin{pmatrix} 1 & 1 \\ 0 & 1 \end{pmatrix} = \begin{pmatrix} 2 & -1 \\ 3 & 4 \\ -1 & 1 \end{pmatrix}$$

from Corollary 2.3.10. ◇

Finding a good base change

Let A and B be in $M_{m \times n}(R)$. Remember that we say A and B are *equivalent* if $B = PAQ$ for some invertible matrices P and Q. (See Definition 2.3.11.) Two matrices are equivalent if and only if they can represent the same linear maps.

Obviously, equivalent matrices should have the same properties. Linear maps which can be represented by the same matrix should have the same properties. To study a linear map, it makes sense to find bases for the domain and codomain so that the corresponding matrix is as simple as possible.

Exercise 10, §3.3, gives us the following result.

Corollary 4.1.3. *Let $A \in M_{m \times n}(F)$ where F is a field. Then A is equivalent to $\begin{pmatrix} I_r & 0 \\ 0 & 0 \end{pmatrix}$ where $r = \mathrm{rk}\, A$.*

Corollary 3.5.8 gives us the following result.

Corollary 4.1.4. *Equivalent matrices over a ring have the same minor ideals. Therefore they are of the same rank.*

Example 4.1.5. Consider the $\mathbb{R}$-linear map T in Example 4.1.2. Find bases for the domain and the codomain of T so that the matrix representing T is of the form in Corollary 4.1.3.

Solution. Let α and α' be the ordered standard bases for $\mathbb{R}^2$ and $\mathbb{R}^3$ respectively. We will give two different methods for finding ordered bases β and β' so that the matrix representing T with respect to β and β' is as required.

Method I. We will apply various elementary matrix operations to

$$A = \begin{pmatrix} 2 & 3 \\ 1 & -1 \\ 5 & -2 \end{pmatrix}$$

so that it becomes the form as in Corollary 4.1.3. First we switch row 1 and row 2:

$$\begin{pmatrix} 0 & 1 & 0 \\ 1 & 0 & 0 \\ 0 & 0 & 1 \end{pmatrix} \begin{pmatrix} 2 & 3 \\ 1 & -1 \\ 5 & -2 \end{pmatrix} = \begin{pmatrix} 1 & -1 \\ 2 & 3 \\ 5 & -2 \end{pmatrix}.$$

Add column 1 to column 2:

$$\begin{pmatrix} 1 & -1 \\ 2 & 3 \\ 5 & -2 \end{pmatrix} \begin{pmatrix} 1 & 1 \\ 0 & 1 \end{pmatrix} = \begin{pmatrix} 1 & 0 \\ 2 & 5 \\ 5 & 3 \end{pmatrix}.$$

Next add $(-2)\times$row 1 to row 2 and then add $(-5)\times$row 1 to row 3:

$$\begin{pmatrix} 1 & 0 & 0 \\ 0 & 1 & 0 \\ -5 & 0 & 1 \end{pmatrix} \begin{pmatrix} 1 & 0 & 0 \\ -2 & 1 & 0 \\ 0 & 0 & 1 \end{pmatrix} \begin{pmatrix} 1 & 0 \\ 2 & 5 \\ 5 & 3 \end{pmatrix} = \begin{pmatrix} 1 & 0 \\ 0 & 5 \\ 0 & 3 \end{pmatrix}.$$

Multiply row 2 by $1/5$:

$$\begin{pmatrix} 1 & 0 & 0 \\ 0 & 1/5 & 0 \\ 0 & 0 & 1 \end{pmatrix} \begin{pmatrix} 1 & 0 \\ 0 & 5 \\ 0 & 3 \end{pmatrix} = \begin{pmatrix} 1 & 0 \\ 0 & 1 \\ 0 & 3 \end{pmatrix}.$$

Finally, add $(-3) \times$ row 2 to row 3:

$$\begin{pmatrix} 1 & 0 & 0 \\ 0 & 1 & 0 \\ 0 & -3 & 1 \end{pmatrix} \begin{pmatrix} 1 & 0 \\ 0 & 1 \\ 0 & 3 \end{pmatrix} = \begin{pmatrix} 1 & 0 \\ 0 & 1 \\ 0 & 0 \end{pmatrix}.$$

Combining all the steps above, we have

$$Q^{-1} \begin{pmatrix} 2 & 3 \\ 1 & -1 \\ 5 & -2 \end{pmatrix} P = \begin{pmatrix} 1 & 0 \\ 0 & 1 \\ 0 & 0 \end{pmatrix}$$

where

$$P = \begin{pmatrix} 1 & 1 \\ 0 & 1 \end{pmatrix}$$

and Q^{-1} is the matrix

$$\begin{pmatrix} 1 & 0 & 0 \\ 0 & 1 & 0 \\ 0 & -3 & 1 \end{pmatrix} \begin{pmatrix} 1 & 0 & 0 \\ 0 & 1/5 & 0 \\ 0 & 0 & 1 \end{pmatrix} \begin{pmatrix} 1 & 0 & 0 \\ 0 & 1 & 0 \\ -5 & 0 & 1 \end{pmatrix} \begin{pmatrix} 1 & 0 & 0 \\ -2 & 1 & 0 \\ 0 & 0 & 1 \end{pmatrix} \begin{pmatrix} 0 & 1 & 0 \\ 1 & 0 & 0 \\ 0 & 0 & 1 \end{pmatrix}.$$

We may view the matrix P as the base change from β to α. Hence we may choose

$$\beta = \left(\begin{pmatrix} 1 \\ 0 \end{pmatrix}, \begin{pmatrix} 1 \\ 1 \end{pmatrix} \right).$$

We may view the matrix Q as the base change matrix from β' to α'. It is clear that Q is the product of elementary matrices

$$\begin{pmatrix} 0 & 1 & 0 \\ 1 & 0 & 0 \\ 0 & 0 & 1 \end{pmatrix} \begin{pmatrix} 1 & 0 & 0 \\ 2 & 1 & 0 \\ 0 & 0 & 1 \end{pmatrix} \begin{pmatrix} 1 & 0 & 0 \\ 0 & 1 & 0 \\ 5 & 0 & 1 \end{pmatrix} \begin{pmatrix} 1 & 0 & 0 \\ 0 & 5 & 0 \\ 0 & 0 & 1 \end{pmatrix} \begin{pmatrix} 1 & 0 & 0 \\ 0 & 1 & 0 \\ 0 & 3 & 1 \end{pmatrix}$$

$$= \begin{pmatrix} 2 & 5 & 0 \\ 1 & 0 & 0 \\ 5 & 3 & 1 \end{pmatrix}.$$

Thus we may choose

$$\beta' = \left(\begin{pmatrix} 2 \\ 1 \\ 5 \end{pmatrix}, \begin{pmatrix} 5 \\ 0 \\ 3 \end{pmatrix}, \begin{pmatrix} 0 \\ 0 \\ 1 \end{pmatrix} \right).$$

Method II. The concept of this method may be used to prove Corollary 4.1.3 (or Exercise 10, §3.3).

Note that $T(e_1)$ and $T(e_2)$ are linearly independent over $\mathbb{R}$. (Incidentally $\operatorname{rk} T = 2$.) We will expand $\{T(e_1),\ T(e_2)\}$ to a basis of $\mathbb{R}^3$. For example, we may choose

$$\beta' = \left(T(e_1) = \begin{pmatrix} 2 \\ 1 \\ 5 \end{pmatrix}, \ T(e_2) = \begin{pmatrix} 3 \\ -1 \\ -2 \end{pmatrix}, \ \begin{pmatrix} 1 \\ 0 \\ 0 \end{pmatrix} \right)$$

since

$$\det \begin{pmatrix} 2 & 3 & 1 \\ 1 & -1 & 0 \\ 5 & -2 & 0 \end{pmatrix} \neq 0.$$

Let $\beta = \alpha$. Then the matrix representing T with respect to β and β' is

$$\begin{pmatrix} 1 & 0 \\ 0 & 1 \\ 0 & 0 \end{pmatrix}.$$

Method II is less tedious if one is comfortable with expanding a linearly independent set to a basis. ◇

In this subsection, we have demonstrated how the rank of a matrix with entries in a field can give us a most "simple and straightforward" equivalent matrix thereof. We also discussed how to find base change matrices for that equivalent matrix.

Exercises 4.1

1. Let R be a ring and let A be an $n \times n$ matrix over R. Show that A is invertible if and only if the columns of A form a basis for R^n over R if and only if the rows of A form a basis for R^n over R.

2. Let $v_1 = (1, 2, 3)$, $v_2 = (1, 0, 2)$ and $v_3 = (1, -1, 2)$. Determine whether v_1, v_2, v_3 form a basis for R^3 over R when $R = \mathbb{Q}$ and $R = \mathbb{Z}$ respectively.

3. Let $v_1 = (1, 2, 3)$, $v_2 = (-1, 0, 2)$ and $v_3 = (0, 3, -2)$. Determine whether v_1, v_2, v_3 form a basis for R^3 over R when $R = \mathbb{Q}$ and $R = \mathbb{Z}$ respectively.

4. Let

$$A = \begin{pmatrix} 1 & 1 & 1 & 1 & 1 \\ 1 & 2 & 3 & 1 & 2 \\ 1 & 2 & 3 & 2 & 1 \\ 1 & 2 & 1 & 1 & 1 \end{pmatrix}$$

be a matrix over $\mathbb{Z}_5$.

   (a) Find $r = \operatorname{rk} A$.

   (b) Find invertible matrices P and Q of appropriate sizes over $\mathbb{Z}_5$ so that $Q^{-1}AP = \begin{pmatrix} \mathbf{I}_r & \mathbf{0} \\ \mathbf{0} & \mathbf{0} \end{pmatrix}$.

4.2   The normal form

Corollary 4.1.3 gives us a very neat result when a matrix has entries in a field. Obviously, we cannot expect this to happen for matrices over an arbitrary ring. However, when the entries of a matrix are in a PID, the situation is quite good as well.

PID's, Euclidean domains and UFD's

Before we continue the discussion, let's first review a few elementary facts in ring theory.

Definition 4.2.1. A **principal ideal domain**, or simply a **PID**, is an integral domain in which every ideal is principal.

The author would like to remind the readers that in this book, the set of *natural numbers* $\mathbb{N}$ is the set of non-negative integers $\{0, 1, 2, 3, \ldots\}$.

Definition 4.2.2. Let D be an integral domain. A **Euclidean valuation** is a function $\nu \colon D \setminus \{0\} \to \mathbb{N}$ such that for any a, $b \in D \setminus \{0\}$, there exist q, $r \in D$ such that

$$a = bq + r, \qquad \text{where } r = 0 \text{ or } \nu(r) < \nu(b).$$

The process of finding q and r for a and b is called the **division algorithm**.

A **Euclidean domain** is an integral domain which can be endowed with a Euclidean valuation.

Remember that any Euclidean domain is a PID. The ring of integers $\mathbb{Z}$, the ring of Gaussian integers $\mathbb{Z}[i]$ and the polynomial ring $F[x]$ of one variable over the field F are some of the best-known examples of Euclidean domains. In the ring $\mathbb{Z}$ we may choose the *absolute value* to be the Euclidean valuation. In the ring $\mathbb{Z}[i]$ we may use the *norm* of a Gaussian integer, while in $F[x]$, we may use the *degree* of a polynomial as the Euclidean valuation.

Another important fact is that a PID is a UFD. Let's review some notions regarding or related to a UFD.

Let D be a domain and let a, $b \in D$. We say a **divides** b or a is a **divisor** of b, denoted $a|b$, if $b = ac$ for some $c \in D$. We say b is an **associate** of a if $b = ua$ for some unit u in D. It is easy to see that being an associate is an equivalence relation. We will write $a \sim b$ for the fact that a and b are associates with each other. Suppose a is nontrivial and non-unit in the domain D.

- We say a is **irreducible** in D if "$a = bc \Rightarrow$ either b or c is a unit in D."

- We say a is **prime** in D if "$a|bc \Rightarrow a|b$ or $a|c$ in D."

Definition 4.2.3. We say a domain D is a **unique factorization domain**, or a **UFD**, or a **factorial** domain, if the following two conditions are satisfied for any $a \in D \setminus \{0\}$:

- *Factorization exists.* The element $a = up_1p_2 \cdots p_s$ where u is a unit and p_i is irreducible for all i.

- *Factorization is unique.* Suppose $a = up_1p_2 \cdots p_s = vq_1q_2 \cdots q_t$ where u, v are units and p_i, q_j are irreducible for all i and j. Then $s = t$ and $p_i \sim q_i$ for all i after a re-ordering of the q_j's.

In a domain, a prime element is always irreducible, while an irreducible element in not necessarily prime. However, in a UFD, any irreducible element is also prime.

Definition 4.2.4. Let S be a subset of an integral domain D. We say $d \in D$ is a **common divisor** of elements in S if d divides every element in S. We say d is a **greatest common divisor**, or simply a **g. c. d.**, of S if whenever we have a common divisor e of S, we have that e divides d.

In a domain, the g. c. d. of a collection of elements does not necessarily exist, and when it exists, it may not be unique. However, any two such g. c. d.'s will be associates to each other. Hence the g. c. d. is *unique* up to associates in a domain. If d is a g. c. d. of a_1, a_2, ..., a_n in a domain D, we write $(a_1, a_2, \ldots, a_n) \sim d$.

In a UFD, every subset of elements must have a g. c. d. This is a strong property of the UFD's.

Let I be an ideal in a PID. One useful fact is that I is generated by a g. c. d. of the elements in I. In fact, I is generated by a g. c. d. of the elements in a generating set for I.

Definition 4.2.5. In a UFD, we say that two elements a and b are **relatively prime** with each other if $(a, b) \sim 1$.

In particular, if $(a, b) \sim 1$ in a PID D, we may find c, d in D such that $ca + db = 1$.

Equivalence of matrices with entries in a PID

When a matrix has entries in a PID, it may be seen to be equivalent to a *diagonal* matrix. The diagonal entries, if carefully chosen, will be of significance to the matrix.

Theorem 4.2.6. *Let $A \in M_{m \times n}(D)$ where D is a PID. We may find nonzero elements $d_1, d_2, \ldots, d_r$ in D where $d_i | d_j$ if $i \leq j$ such that A is equivalent to*

$$(4.2.1) \qquad \mathrm{diag}\,\{d_1, d_2, \ldots, d_r, 0, \ldots, 0\} = \begin{pmatrix} d_1 & & & & & \\ & \ddots & & & & \\ & & d_r & & & \\ & & & 0 & & \\ & & & & \ddots & \end{pmatrix}.$$

The matrix $\mathrm{diag}\,\{d_1, d_2, \ldots, d_r, 0, \ldots, 0\}$ *is called a **normal form** of A.*

Before proving this theorem, we will use some examples to demonstrate the idea in its proof. Because these examples are over Euclidean domains, the situation is in fact even simpler. We only need to apply elementary column/row operations to a matrix to reach a normal form.

Observing the normal form in (4.2.1), we can see that the d_1 is a g.c.d. of the entries in the normal form. This implies that d_1 generates the 1-minor ideal of the normal form. By Corollary 3.5.8, d_1 is a generator of $I_1(A)$. Hence the starting point of obtaining the normal form is to produce a g.c.d. as the $(1,1)$-entry using various elementary operations. To achieve this in the case of Euclidean domains, we minimize the Euclidean valuation of the entries until eventually we have a g.c.d. somewhere. And then we move the g.c.d. to its rightful place.

Let's start with a smaller matrix to demonstrate our point.

Example 4.2.7. Obtain a normal form for

$$\begin{pmatrix} 2 & 3 \\ 4 & 5 \end{pmatrix} \in M_2(\mathbb{Z}).$$

Solution. We start by adding $(-1)\times$column 1 to column 2 to produce a g.c.d. of A:

$$A \rightsquigarrow \begin{pmatrix} 2 & \boxed{1} \\ 4 & 1 \end{pmatrix}, \quad \text{producing a g.c.d. as the } (1,2)\text{-entry,}$$

$$\rightsquigarrow \begin{pmatrix} \boxed{1} & 2 \\ 1 & 4 \end{pmatrix}, \quad \text{moving the g.c.d. to the } (1,1)\text{-place.}$$

The g.c.d. is the $(1,1)$-entry now, and we may use it to delete the entries at its column and at its row:

$$\begin{pmatrix} \boxed{1} & 2 \\ 1 & 4 \end{pmatrix} \rightsquigarrow \begin{pmatrix} \boxed{1} & 0 \\ 1 & 2 \end{pmatrix}, \quad \text{deleting the other entry at row 1,}$$

$$\rightsquigarrow \begin{pmatrix} \boxed{1} & 0 \\ 0 & 2 \end{pmatrix}, \quad \text{deleting the other entry at column 1.}$$

Note that $1 \mid 2$ in $\mathbb{Z}$. The matrix diag$\{1, 2\}$ is a normal form for A.   ◇

As we can see from the previous example, our goal is (i) to produce a g.c.d. of entries, (ii) to move the g.c.d. to the upper left corner, and (iii) to use the g.c.d. to cancel entries at is row and at its column.

Example 4.2.8. Obtain a normal form for the matrix

$$A = \begin{pmatrix} 3 & 1 & 3 & 1 & -1 \\ 2 & -6 & -2 & 4 & 2 \\ -3 & 2 & 0 & 1 & 3 \\ 4 & 2 & 6 & -2 & 4 \end{pmatrix} \in M_{4\times 5}(\mathbb{Z}).$$

Solution. The $(1,2)$-entry of A happens to be a g.c.d. of all the entries in A. We first switch column 1 and column 2 so that the $(1,1)$-entry is now the desired g.c.d.:

$$A \rightsquigarrow \begin{pmatrix} \boxed{1} & 3 & 3 & 1 & -1 \\ -6 & 2 & -2 & 4 & 2 \\ 2 & -3 & 0 & 1 & 3 \\ 2 & 4 & 6 & -2 & 4 \end{pmatrix}$$

$$\rightsquigarrow \begin{pmatrix} \boxed{1} & 0 & 0 & 0 & 0 \\ -6 & 20 & 16 & 10 & -4 \\ 2 & -9 & -6 & -1 & 5 \\ 2 & -2 & 0 & -4 & 6 \end{pmatrix}, \quad \text{deleting other entries in row 1,}$$

$$\rightsquigarrow \begin{pmatrix} \boxed{1} & 0 & 0 & 0 & 0 \\ 0 & 20 & 16 & 10 & -4 \\ 0 & -9 & -6 & -1 & 5 \\ 0 & -2 & 0 & -4 & 6 \end{pmatrix}, \quad \text{deleting other entries in column 1.}$$

Before continuing, double check that the $(1,1)$-entry is indeed a factor of all the other entries at this stage. We may now concentrate on the 3×4 submatrix at the lower right hand corner. The $(3,4)$-entry happens to be a g.c.d. of the entries in this submatrix. We will first change the $(3,4)$-entry to 1 by multiplying the fourth column by -1, but this step is optional. We will then move it the $(2,2)$-place. Clearly the $(1,1)$-entry will divide the $(2,2)$-entry since it is a common divisor of all the entries in the original 3×4 submatrix. We have that

$$A \dashrightarrow \begin{pmatrix} 1 & 0 & 0 & 0 & 0 \\ 0 & 20 & 16 & -10 & -4 \\ 0 & -9 & -6 & \boxed{1} & 5 \\ 0 & -2 & 0 & 4 & 6 \end{pmatrix}$$

$$\leadsto \begin{pmatrix} 1 & 0 & 0 & 0 & 0 \\ 0 & \boxed{1} & -6 & -9 & 5 \\ 0 & -10 & 16 & 20 & -4 \\ 0 & 4 & 0 & -2 & 6 \end{pmatrix}, \quad \text{moving the g.\,c.\,d.,}$$

$$\leadsto \begin{pmatrix} 1 & 0 & 0 & 0 & 0 \\ 0 & \boxed{1} & -6 & -9 & 5 \\ 0 & 0 & -44 & -70 & 46 \\ 0 & 0 & 24 & 34 & -14 \end{pmatrix}, \quad \text{deleting entries,}$$

$$\leadsto \begin{pmatrix} 1 & 0 & 0 & 0 & 0 \\ 0 & \boxed{1} & 0 & 0 & 0 \\ 0 & 0 & -44 & -70 & 46 \\ 0 & 0 & 24 & 34 & -14 \end{pmatrix}, \quad \text{deleting entries.}$$

Next we concentrate on transforming the 2×3 submatrix at the lower right hand corner. None of the entries is a g.c.d. in that submatrix. Note that $(4,5)$-entry has the least (nonzero) Euclidean valuation (the absolute value in this case). We may add $2\times$column 5 to column 4 using the division algorithm to obtain 6 as an entry, which is less than the original element of minimal valuation. In actual practice, one may notice that we can obtain 2 as an entry by adding column 3 to column 5. Since $|2| < |6|$, this speeds up the calculation process. Note that 2 is also a g.c.d. of the entries in the 2×3 submatrix at the lower right hand corner. Now we have

$$A \dashrightarrow \begin{pmatrix} 1 & 0 & 0 & 0 & 0 \\ 0 & 1 & 0 & 0 & 0 \\ 0 & 0 & -44 & -70 & \boxed{2} \\ 0 & 0 & 24 & 34 & 10 \end{pmatrix}$$

$$\leadsto \begin{pmatrix} 1 & 0 & 0 & 0 & 0 \\ 0 & 1 & 0 & 0 & 0 \\ 0 & 0 & \boxed{2} & -70 & -44 \\ 0 & 0 & 10 & 34 & 24 \end{pmatrix}, \quad \text{moving the g.\,c.\,d.,}$$

$$\leadsto \begin{pmatrix} 1 & 0 & 0 & 0 & 0 \\ 0 & 1 & 0 & 0 & 0 \\ 0 & 0 & \boxed{2} & -70 & -44 \\ 0 & 0 & 0 & 384 & 244 \end{pmatrix}, \quad \text{deleting entries,}$$

$$\rightsquigarrow \begin{pmatrix} 1 & 0 & 0 & 0 & 0 \\ 0 & 1 & 0 & 0 & 0 \\ 0 & 0 & \boxed{2} & 0 & 0 \\ 0 & 0 & 0 & 384 & 244 \end{pmatrix}, \quad \text{deleting entries.}$$

We are now at the final 1×2 submatrix. We will use the division algorithm to keep reducing the absolute value of the entries in this submatrix:

$$A \dashrightarrow \begin{pmatrix} 1 & 0 & 0 & 0 & 0 \\ 0 & 1 & 0 & 0 & 0 \\ 0 & 0 & 2 & 0 & 0 \\ 0 & 0 & 0 & 140 & 244 \end{pmatrix}, \quad 140 = 384 - 244,$$

$$\rightsquigarrow \begin{pmatrix} 1 & 0 & 0 & 0 & 0 \\ 0 & 1 & 0 & 0 & 0 \\ 0 & 0 & 2 & 0 & 0 \\ 0 & 0 & 0 & 140 & 104 \end{pmatrix}, \quad 104 = 244 - 140,$$

$$\rightsquigarrow \begin{pmatrix} 1 & 0 & 0 & 0 & 0 \\ 0 & 1 & 0 & 0 & 0 \\ 0 & 0 & 2 & 0 & 0 \\ 0 & 0 & 0 & 36 & 104 \end{pmatrix}, \quad 36 = 140 - 104,$$

$$\rightsquigarrow \begin{pmatrix} 1 & 0 & 0 & 0 & 0 \\ 0 & 1 & 0 & 0 & 0 \\ 0 & 0 & 2 & 0 & 0 \\ 0 & 0 & 0 & 36 & \boxed{-4} \end{pmatrix}, \quad -4 = 104 - 36 \cdot 3.$$

We now have reached a g. c. d. for the final 1×2 submatrix. The final step is to move the g. c. d. to column 4 and use it to delete the other nonzero entry in row 4. We will also adjust the sign of the g. c. d.:

$$A \dashrightarrow \begin{pmatrix} 1 & 0 & 0 & 0 & 0 \\ 0 & 1 & 0 & 0 & 0 \\ 0 & 0 & 2 & 0 & 0 \\ 0 & 0 & 0 & \boxed{-4} & 36 \end{pmatrix} \rightsquigarrow \begin{pmatrix} 1 & 0 & 0 & 0 & 0 \\ 0 & 1 & 0 & 0 & 0 \\ 0 & 0 & 2 & 0 & 0 \\ 0 & 0 & 0 & \boxed{4} & 0 \end{pmatrix}.$$

The matrix diag $\{1, 1, 2, 4\}$ is a normal form for A.   ◇

The following example involves a matrix with entries in $\mathbb{Z}[i]$, the ring of Gaussian integers. We may use the *norm* in $\mathbb{Z}[i]$ to be the Euclidean valuation. Remember that the norm

$$N(\alpha) = \alpha\overline{\alpha} = a^2 + b^2$$

for $\alpha = a + bi$ where $a, b \in \mathbb{Z}$.

Example 4.2.9. Obtain a normal form for the matrix

$$A = \begin{pmatrix} 1 & 3 & 6 \\ 2+3i & -3i & 12-18i \\ 2-3i & 6+9i & -18i \end{pmatrix}$$

over $\mathbb{Z}[i]$.

Solution. The $(1,1)$-entry of A is already a g.c.d. of all the entries in A. We use it to delete all other entries in row 1 and column 1:

$$A \rightsquigarrow \begin{pmatrix} \boxed{1} & 0 & 0 \\ 2+3i & -6-12i & -36i \\ 2-3i & 18i & -12 \end{pmatrix} \rightsquigarrow \begin{pmatrix} \boxed{1} & 0 & 0 \\ 0 & -6-12i & -36i \\ 0 & 18i & -12 \end{pmatrix}.$$

Next we use the division algorithm to produce a g.c.d. for the entries in the lower right 2×2 submatrix:

$$A \dashrightarrow \begin{pmatrix} 1 & 0 & 0 \\ 0 & 30-12i & -36i \\ 0 & \boxed{6i} & -12 \end{pmatrix}, \quad 18i = 12(i) + 6i,$$

$$\rightsquigarrow \begin{pmatrix} 1 & 0 & 0 \\ 0 & \boxed{6i} & -12 \\ 0 & 30-12i & -36i \end{pmatrix}, \quad \text{moving the g.c.d.,}$$

$$\rightsquigarrow \begin{pmatrix} 1 & 0 & 0 \\ 0 & \boxed{6} & 12i \\ 0 & 30-12i & -36i \end{pmatrix}, \quad \text{adjusting the g.c.d. (optional),}$$

$$\rightsquigarrow \begin{pmatrix} 1 & 0 & 0 \\ 0 & \boxed{6} & 0 \\ 0 & 30-12i & -24-96i \end{pmatrix}, \quad \text{deleting entries,}$$

$$\rightsquigarrow \begin{pmatrix} 1 & 0 & 0 \\ 0 & \boxed{6} & 0 \\ 0 & 0 & -24-96i \end{pmatrix}, \quad \text{deleting entries,}$$

$$\rightsquigarrow \begin{pmatrix} 1 & 0 & 0 \\ 0 & 6 & 0 \\ 0 & 0 & \boxed{24+96i} \end{pmatrix}, \quad \text{adjusting the g.\,c.\,d. (optional).}$$

The matrix diag $\{1, 6, 24 + 96i\}$ is a normal form for A.   ◇

The next example involves a matrix with entries in the polynomial ring of one variable over $\mathbb{Q}$. We will use the *degree* of a nonzero polynomial as the Euclidean valuation for this polynomial ring.

Example 4.2.10. Let λ be an indeterminate over $\mathbb{Q}$. Obtain a normal form of the matrix

$$A = \begin{pmatrix} \lambda - 17 & 8 & 12 & -14 \\ -46 & \lambda + 22 & 35 & -41 \\ 2 & -1 & \lambda - 4 & 4 \\ -4 & 2 & 2 & \lambda - 3 \end{pmatrix}$$

over $\mathbb{Q}[\lambda]$.

The calculation in this example is more complicated than in the previous examples. We will adjust the steps as we see fit.

Solution. The g.\,c.\,d. of the entries of A is obviously 1. We will start with the $(3, 2)$-entry. We use it to cancel the other entries in column 2 and in row 3:

$$A \rightsquigarrow \begin{pmatrix} \lambda - 1 & 0 & 8\lambda - 20 & 18 \\ 2\lambda - 2 & 0 & \lambda^2 + 18\lambda - 53 & 4\lambda + 47 \\ 2 & \boxed{-1} & \lambda - 4 & 4 \\ 0 & 0 & 2\lambda - 6 & \lambda + 5 \end{pmatrix}$$

$$\rightsquigarrow \begin{pmatrix} \lambda - 1 & 0 & 8\lambda - 20 & 18 \\ 2\lambda - 2 & 0 & \lambda^2 + 18\lambda - 53 & 4\lambda + 47 \\ 0 & \boxed{-1} & 0 & 0 \\ 0 & 0 & 2\lambda - 6 & \lambda + 5 \end{pmatrix}.$$

We then move and adjust the g.\,c.\,d. to its rightful place:

$$A \dashrightarrow \begin{pmatrix} \boxed{1} & 0 & 0 & 0 \\ 0 & \lambda - 1 & 8\lambda - 20 & 18 \\ 0 & 2\lambda - 2 & \lambda^2 + 18\lambda - 53 & 4\lambda + 47 \\ 0 & 0 & 2\lambda - 6 & \lambda + 5 \end{pmatrix}.$$

Next we will concentrate on "normalizing" the 3×3 submatrix at the lower right hand corner. Observe the g.c.d. of this submatrix is again 1 (the $(2, 4)$-entry being a unit). We should be able to produce the number 1 as the $(2, 2)$-entry. At this point, there are many ways to handle this matrix. Instead of moving the $(2, 4)$-entry to the $(2, 2)$-place, we choose to add $(-2) \times$row 2 to row 3. This is simply to avoid unnecessarily tedious computation involving too many non-integers while producing more zero entries in the matrix at the same time. We have

$$A \sim \begin{pmatrix} 1 & 0 & 0 & 0 \\ 0 & \lambda - 1 & 8\lambda - 20 & 18 \\ 0 & 0 & \lambda^2 + 2\lambda - 13 & 4\lambda + 11 \\ 0 & 0 & 2\lambda - 6 & \lambda + 5 \end{pmatrix}.$$

Next we add $(-8) \times$column 2 to column 3. Here we are using the division algorithm to reduce the Euclidean valuation of the $(2, 3)$-entry. We have

$$A \sim \begin{pmatrix} 1 & 0 & 0 & 0 \\ 0 & \lambda - 1 & -12 & 18 \\ 0 & 0 & \lambda^2 + 2\lambda - 13 & 4\lambda + 11 \\ 0 & 0 & 2\lambda - 6 & \lambda + 5 \end{pmatrix}.$$

The number -12 of the $(2, 3)$-entry still feels troublesome. Add column 4 to column 3 and we have

$$A \sim \begin{pmatrix} 1 & 0 & 0 & 0 \\ 0 & \lambda - 1 & 6 & 18 \\ 0 & 0 & \lambda^2 + 6\lambda - 2 & 4\lambda + 11 \\ 0 & 0 & 3\lambda - 1 & \lambda + 5 \end{pmatrix}.$$

Multiply column 2 by 6. This is permissible since 6 is a unit in $\mathbb{Q}[\lambda]$. We have

$$A \sim \begin{pmatrix} 1 & 0 & 0 & 0 \\ 0 & 6\lambda - 6 & 6 & 18 \\ 0 & 0 & \lambda^2 + 6\lambda - 2 & 4\lambda + 11 \\ 0 & 0 & 3\lambda - 1 & \lambda + 5 \end{pmatrix}.$$

Add $-(\lambda - 1) \times$column 3 to column 2 and $(-3) \times$column 3 to column 4 to obtain

$$A \sim \begin{pmatrix} 1 & 0 & 0 & 0 \\ 0 & 0 & \boxed{6} & 0 \\ 0 & -(\lambda^3 + 5\lambda^2 - 8\lambda + 2) & \lambda^2 + 6\lambda - 2 & -(3\lambda^2 + 14\lambda - 17) \\ 0 & -(3\lambda^2 - 4\lambda + 1) & 3\lambda - 1 & -(8\lambda - 8) \end{pmatrix}.$$

Multiply columns 2 and 4 by -1 to make the leading coefficients of the nonzero terms in these columns positive. Multiply row 2 by $1/6$ to adjust the g. c. d. at the $(2,3)$-place. We then use the g. c. d. to delete all other entries in column 3. Finally we move the $(2,3)$-entry to the $(2,2)$-place. Now we have

$$A \sim \begin{pmatrix} 1 & 0 & 0 & 0 \\ 0 & \boxed{1} & 0 & 0 \\ 0 & 0 & (\lambda - 1)(\lambda^2 + 6\lambda - 2) & (\lambda - 1)(3\lambda + 17) \\ 0 & 0 & (\lambda - 1)(3\lambda - 1) & 8(\lambda - 1) \end{pmatrix}.$$

The g. c. d. of the entries in the 2×2 submatrix at the lower right hand corner is $\lambda - 1$, which will be the $(3,3)$-entry of a normal form for A. First multiply column 3 by 8. Next add $-(3\lambda - 1) \times$column 4 to column 3 to obtain

$$A \sim \begin{pmatrix} 1 & 0 & 0 & 0 \\ 0 & 1 & 0 & 0 \\ 0 & 0 & 8(\lambda - 1)(\lambda^2 + 6\lambda - 2) & (\lambda - 1)(3\lambda + 17) \\ 0 & 0 & 8(\lambda - 1)(3\lambda - 1) & 8(\lambda - 1) \end{pmatrix}$$

$$\sim \begin{pmatrix} 1 & 0 & 0 & 0 \\ 0 & 1 & 0 & 0 \\ 0 & 0 & (\lambda - 1)(\lambda^2 - 1) & (\lambda - 1)(3\lambda + 17) \\ 0 & 0 & 0 & \boxed{8(\lambda - 1)} \end{pmatrix}.$$

Multiply row 4 by $1/8$ so that the $(4,4)$-entry becomes $\lambda - 1$. We may use the $(4,4)$-entry to delete the other entries in column 4. We then move it to its rightful place, the $(3,3)$-place. We have

$$A \sim \begin{pmatrix} 1 & 0 & 0 & 0 \\ 0 & 1 & 0 & 0 \\ 0 & 0 & \boxed{\lambda - 1} & 0 \\ 0 & 0 & 0 & \boxed{(\lambda - 1)(\lambda^2 - 1)} \end{pmatrix}.$$

This is a normal form for A.   ◇

Hopefully, the examples above help to demonstrate the process of "normalizing" a matrix, at least for the case of Euclidean domains. In the following proof we are simply describing the process in generalities.

Proof of Theorem 4.2.6. We may assume the matrix A in question is nontrivial without loss of generality. We will prove the theorem by induction on m, the number of rows of A.

Case 1. Assume D is a Euclidean domain. Let $\nu \colon D \setminus \{0\} \to \mathbb{N}$ be a Euclidean valuation on D.

Locate an entry with the least Euclidean valuation. Move that entry to the $(1,1)$-place by rotating rows and/or columns. Observe whether the new $(1,1)$-entry, say a, divides all other entries in the first row and in the first column. If it happens that a does not divide, say a_{1i}, for some i, find q and $r \in D$ such that $a_{1i} = aq + r$ and $\nu(r) < \nu(a)$. Apply an elementary column operation of type III by adding $(-q) \times$ the first column to the i-th column, and the matrix A is equivalent to a matrix of the form

$$\begin{pmatrix} a & \cdots & r & \cdots \\ & & & \\ * & & \ast & \\ & & & \end{pmatrix}.$$

Switch the first column and the i-th column and now the $(1,1)$-entry is r. We have successfully reduced the Euclidean valuation of the $(1,1)$-entry. If at this point the $(1,1)$-entry still does not divide all other entries in the first column or in the first row, we may repeat the process above to further reduce the Euclidean valuation. Since the Euclidean valuation cannot be reduced indefinitely, eventually the $(1,1)$-entry must divide all other entries in the first column and in the first row. We may now use the $(1,1)$-entry to cancel all other entries in its column and in its row so that the matrix is of the form

(4.2.2)
$$\begin{pmatrix} b & 0 & \cdots & 0 \\ 0 & & & \\ \vdots & & \ast & \\ 0 & & & \end{pmatrix}.$$

Next we search whether there are other entries in the matrix not divisible by b. Suppose the (i, j)-entry is not divisible by b. Add the i-th row to the first row to produce the said entry in the first row. We then repeat the process in the last paragraph until the matrix is back to the form as in (4.2.2). Repeat the process until the matrix finally reaches the form

(4.2.3)
$$B = \begin{pmatrix} d_1 & 0 & \cdots & 0 \\ 0 & & & \\ \vdots & & C & \\ 0 & & & \end{pmatrix}$$

where d_1 divides all entries in the lower right hand corner.

Note that d_1 is a g. c. d. of all the entries in B, which is also a generator for the ideal generated by the 1-minors of B since D is a PID. By Corollary 3.5.8, the ideal generated by the 1-minors of A is the same as the ideal generated by the 1-minors of B. Hence d_1 is also a g. c. d. of the entries in A.

Observe that the matrix C has one less rows than A. By induction on the number of rows we have that C is equivalent to diag $\{d_2, \ldots, d_r, 0, \ldots, 0\}$ where $d_i | d_j$ for $2 \leq i \leq j \leq r$. More precisely, we may find invertible matrices U and V such that

$$UCV = \text{diag}\{d_2, \ldots, d_r, 0, \ldots, 0\}.$$

From the previous discussion we can see that d_2 is a g. c. d. of all the entries in C. Since d_1 also divides every entries in C, we have that d_1 divides d_2. Thus A is equivalent to

$$\begin{pmatrix} 1 & 0 & \cdots & 0 \\ 0 & & & \\ \vdots & & U & \\ 0 & & & \end{pmatrix} \begin{pmatrix} d_1 & 0 & \cdots & 0 \\ 0 & & & \\ \vdots & & C & \\ 0 & & & \end{pmatrix} \begin{pmatrix} 1 & 0 & \cdots & 0 \\ 0 & & & \\ \vdots & & V & \\ 0 & & & \end{pmatrix}$$

$$= \text{diag}\{d_1, d_2, \ldots, d_r, 0, \ldots, 0\}$$

where $d_i | d_j$ for $1 \leq i \leq j \leq r$.

Case 2. Assume that D is a PID. The argument is quite similar to that of Case 1. The only difference is that we will use the *length* of a nonzero element in D to replace the Euclidean valuation in Case 1.

Let $a \in D \setminus \{0\}$. Define the *length* of a, denoted $\ell(a)$, to be the number of prime factors occurring in a factorization of a into irreducibles. To be more clear, $\ell(a) = 0$ if a is a unit in D, and $\ell(a) = r$ if $a = up_1 p_2 \cdots p_r$ where u is a unit and the p_i's are primes in D.

First, find an entry of least length in the matrix A and move it to the $(1,1)$-place. Suppose in the first row there is an element b such that $a \nmid b$. Switch the second column and the column of b if necessary so that we may assume the first row of the matrix to be $(a, b, *, \ldots, *)$. Let d be a g.c.d. of a and b. Note that $\ell(d) < \ell(a)$. Since D is a PID, the ideal $(a, b) = (d)$. We may find x and y in D such that $ax + by = d$. Let

$$
X = \begin{pmatrix}
x & b/d & & & & 0 \\
y & -a/d & & & & \\
& & 1 & & & \\
& & & 1 & & \\
0 & & & & \ddots & \\
& & & & & 1
\end{pmatrix}.
$$

Since $\det X = (-ax - by)/d = -1$, it is invertible by Theorem 3.4.9. Multiply our matrix by X on the right, we can see that the matrix A is equivalent to a matrix whose first row is of the form $(d, 0, *, \ldots, *)$. Similarly, if b is an element in the first column such that $a \nmid b$, we may switch rows so that the first column is of the form $(a, b, *, \ldots, *)^{\mathrm{tr}}$. Multiply the original matrix by X^{tr} on the left and we see that A is equivalent to a matrix whose first column is of the form $(d, 0, *, \ldots, *)^{\mathrm{tr}}$. In either case, we have reduced the length of the $(1,1)$-entry.

The length of the $(1,1)$-entry can not be reduced indefinitely. Eventually the $(1,1)$-entry will divide all other entries in the first row and in the first column. We may then use the $(1,1)$-entry to cancel all other entries both in its row and column. The matrix is now equivalent to a matrix of the form as in (4.2.2). If at this point, the $(1,1)$-entry still does not divide every entry in the matrix, we may produce the "troublesome" entry in the first row or column by using an elementary operation of type III. Repeat the aforementioned process to further reduce the length of the $(1,1)$-entry until our matrix is equivalent to a matrix of the form as in (4.2.3) where d_1 divides every entry in C. The theorem now follows by induction on the number of rows of A using the same argument as in Case 1. $\qquad \square$

Uniqueness of the normal form

Let $A \in M_{m \times n}(D)$ where D is a PID and let $d_1, d_2, \ldots, d_r$ be nonzero elements in D such that

$$B = \text{diag}\{d_1, d_2, \ldots, d_r, 0, \ldots, 0\}$$

is a normal form for A. The *nonzero* diagonal entries of B are called **invariant factors** of A. This terminology would make sense once the uniqueness of the normal form is established. Remember that the minor ideals of A and of B are the same. By Corollary 3.5.8 we have that

$$I_i(A) = I_i(B) = (d_{j_1} d_{j_2} \cdots d_{j_i} : 1 \leq j_1 < j_2 < \cdots < j_i \leq r)$$

for all $1 \leq i \leq r$. Since D is a PID, each of the minor ideals is generated by a g.c.d. of its generators. It follows that

$$I_i(A) = I_i(B) = \begin{cases} (d_1 d_2 \cdots d_i), & \text{for } 1 \leq i \leq r; \\ (0), & \text{for } i > r, \end{cases}$$

since $d_i | d_j$ for $i < j$. Thus the invariant factors of A can be recovered from the minor ideals of A. We now have the following result.

Theorem 4.2.11. *Let A be an $m \times n$ matrix with entries in an PID and suppose that $\text{rk } A = r$. For each $i \leq r$, let Δ_i be a g.c.d. of all the i-minors of A. Then the invariant factors of A are*

$$d_1 \sim \Delta_1, \ d_2 \sim \Delta_2 \Delta_1^{-1}, \ \ldots, \ d_r \sim \Delta_r \Delta_{r-1}^{-1}$$

up to associates in D.

Remember that $(0) \subsetneqq I_r(A) \subsetneqq \cdots \subsetneqq I_2(A) \subsetneqq I_1(A)$. Hence we should have $\Delta_i \neq 0$ for $1 \leq i \leq r$ and $\Delta_i | \Delta_{i+1}$ for $1 \leq i \leq r - 1$.

Exercises 4.2

The first two exercises are review problems for UFD's.

1. In a UFD, we say a collection of primes are **distinct** if they are non-associates to each other.

Let D be a UFD. Let $a = up_1^{i_1} p_2^{i_2} \cdots p_s^{i_s}$ and $b = vp_1^{j_1} p_2^{j_2} \cdots p_s^{j_s}$ where u, v are units, the p_i's are distinct prime in D and i_k, $j_k \geq 0$ for all k. Show that $a|b$ if and only if $i_k \leq j_k$ for all k.

2. Let D be a UFD. Suppose a, b, $c \in D$ such that a and b are relatively prime with each other.

   (a) Show that $a|c$ if $a|b^k c$.

   (b) Show that $ab|c$ if $a|c$ and $b|c$.

3. Obtain the normal form of

$$\begin{pmatrix} 6 & 2 & -3 & -1 \\ 2 & 3 & 3 & 2 \\ 3 & -4 & 1 & -3 \\ 0 & 1 & 2 & 5 \end{pmatrix}$$

   over $\mathbb{Z}$.

4. Let λ be an indeterminate over $\mathbb{Q}$. Find the normal form of

$$\begin{pmatrix} \lambda+1 & 2 & -6 \\ 1 & \lambda & -3 \\ 1 & 1 & \lambda-4 \end{pmatrix}$$

   over $\mathbb{Q}[\lambda]$.

5. Let's generalize the following fact to invertible matrices over a Euclidean domain. *Cf.* Proposition 3.3.2.[1]

   Let A be an invertible matrix of size n over a Euclidean domain D. Show that there are elementary matrices E_1, E_2, ..., E_N over D such that $E_1 E_2 \cdots E_N A = I_n$. Hence any invertible matrix over a Euclidean domain is a product of elementary matrices.

6. Let $A \in M_n(F)$ where F is a field.

---

[1]This is not true for PID's.

(a) Let A be an invertible matrix. Show that we may transform A into a diagonal matrix using only elementary row operations of type III.

(b) Let u, v be nonzero elements in F. Show that we may transform $\text{diag}\{\, u,\, v\,\}$ into $\text{diag}\{\, 1,\, uv\,\}$ using only elementary (column or row) operations of type III. (Hint: Show that one has

$$\text{diag}\{\, u,\, v\,\} \sim \begin{pmatrix} u & 1 \\ & v \end{pmatrix} \sim \begin{pmatrix} 1 & 1 \\ v - vu & v \end{pmatrix}$$

using only elementary operations of type III.)

(c) Let $\det A = 1$. Show that A is a product of elementary matrices of type III.

7. Let D be a Euclidean domain and let ν be a Euclidean valuation on D. Let $A \in M_n(D)$ and assume that $\det A \neq 0$. Show that there exists an invertible $P \in M_n(D)$ such that

$$PA = \begin{pmatrix} d_1 & b_{12} & b_{13} & \cdots & b_{1n} \\ & d_2 & b_{23} & \cdots & b_{2n} \\ & & d_3 & \cdots & b_{3n} \\ & & & \ddots & \vdots \\ & & & & d_n \end{pmatrix}$$

where (i) $d_j \neq 0$ for all j, and (ii) if $b_{ji} \neq 0$ then $\nu(b_{ij}) < \nu(d_j)$.

8. Let D be a PID and let $(a_1, a_2, \ldots, a_n) \sim d$ in D. Show that there is an invertible matrix $Q \in M_n(D)$ such that

$$\begin{pmatrix} a_1 & a_2 & \cdots & a_n \end{pmatrix} Q = \begin{pmatrix} d & 0 & \cdots & 0 \end{pmatrix}.$$

9. Let D be a PID and let $(a_{11}, a_{12}, \ldots, a_{1n}) \sim 1$ in D. Show that there exists $a_{ij} \in D$, $2 \leq i \leq n$ and $1 \leq j \leq n$, such that the square matrix (a_{ij}) is invertible in $M_n(D)$. (Hint: Use Exercise 8.)

4.3 Structure theorem of finitely generated modules over a PID

In this section we will discuss how to treat a finitely generated module over a PID as the cokernel of a matrix and how to use the normal form of the said matrix to analyze its structure.

Throughout this section R denotes a ring.

Cyclic modules

Remember that an R-module M is **cyclic** if M can be generated by a single element over R. In this case, there is an element m in M such that $M = Rm$.

Definition 4.3.1. Let M be an R-module and let $m \in M$. Consider the R-linear map

$$
(4.3.1) \qquad
\begin{array}{rrcl}
f: & R & \longrightarrow & M \\
& 1 & \longmapsto & m \\
& r & \longmapsto & rm
\end{array}
$$

The kernel of f is called the **annihilator** of m in R, denoted $\mathrm{ann}_R\, m$ or simply $\mathrm{ann}\, m$. In other words,

$$ \mathrm{ann}_R\, m = \{\, r \in R : rm = 0 \,\}. $$

Being the kernel of an R-linear map, $\mathrm{ann}_R\, m$ is both an submodule and an ideal of R for any $m \in M$.

Lemma 4.3.2. *Let $M = Rm$ be cyclic over R. Then*

$$ M \cong \frac{R}{\mathrm{ann}_R\, m} \qquad and \qquad \mathrm{Ann}_R\, M = \mathrm{ann}_R\, m. $$

Proof. The R-linear map f in (4.3.1) is onto when $M = Rm$. The first result follows from the first isomorphism theorem.

For the second part, note that $\mathrm{Ann}\, M \subseteq \mathrm{ann}\, m$. Suppose $r \in \mathrm{ann}\, m$. For any $a \in R$, $r(am) = a(rm) = 0$. Hence $r \in \mathrm{Ann}\, M$. Thus we also have $\mathrm{ann}\, m \subseteq \mathrm{Ann}\, M$. $\qquad\square$

From Example 1.2.7 and Lemma 4.3.2 we have the following result.

Corollary 4.3.3. *An R-module is cyclic if and only if it is isomorphic to R/I for some ideal I of R.*

Definition 4.3.4. If $M = Rm$ is a cyclic module over R. The ideal $\mathrm{ann}_R m$ is also called the **order ideal** of the cyclic module.

Example 4.3.5. Consider $M = \mathbb{Z} \oplus \mathbb{Z}/\langle(6, 9), (2, 2)\rangle$ as a $\mathbb{Z}$-module. Find $\mathrm{ann}_\mathbb{Z} \overline{(4, 6)}$ and $\mathrm{ann}_\mathbb{Z} \overline{(4, -6)}$.

Solution. Here we will find the annihilators by brutal force. The situation will become more transparent after we have the structure theorem.

To find $\mathrm{ann}_\mathbb{Z} \overline{(4, 6)}$, let $r \in \mathbb{Z}$ and $r\overline{(4, 6)} = 0$. There exist a and $b \in \mathbb{Z}$ such that

$$(4r, 6r) = a(6, 9) + b(2, 2) = (6a + 2b, 9a + 2b)$$
$$\implies 2r = 6r - 4r = 9a + 2b - (6a + 2b) = 3a \implies 3|r.$$

This implies that $\mathrm{ann}_\mathbb{Z} \overline{(4, 6)} \subseteq 3\mathbb{Z}$. On the other hand, we have the relation $3\overline{(4, 6)} = \overline{(12, 18)} = 2\overline{(6, 9)} = 0$, which implies that $3\mathbb{Z} \subseteq \mathrm{ann}_\mathbb{Z} \overline{(4, 6)}$. We conclude that $\mathrm{ann}_\mathbb{Z} \overline{(4, 6)} = 3\mathbb{Z}$.

To find $\mathrm{ann}_\mathbb{Z} \overline{(4, -6)}$, let $r \in \mathbb{Z}$ and $r\overline{(4, -6)} = 0$. There exist a and $b \in \mathbb{Z}$ such that

$$(4r, -6r) = a(6, 9) + b(2, 2) = (6a + 2b, 9a + 2b)$$
$$\implies -10r = -6r - 4r = 9a + 2b - (6a + 2b) = 3a \implies 3|r.$$

This implies that $\mathrm{ann}_\mathbb{Z} \overline{(4, -6)} \subseteq 3\mathbb{Z}$. On the other hand, we have the relation $3\overline{(4, -6)} = \overline{(12, -18)} = -10\overline{(6, 9)} + 36\overline{(2, 2)} = 0$, which implies that $3\mathbb{Z} \subseteq \mathrm{ann}_\mathbb{Z} \overline{(4, -6)}$. We conclude that $\mathrm{ann}_\mathbb{Z} \overline{(4, -6)} = 3\mathbb{Z}$ as well.   ◇

Cokernel

Suppose given a finitely generated R-module

$$M = Rz_1 + Rz_2 + \cdots + Rz_m.$$

We can define a surjective R-linear map

(4.3.2)
$$\begin{array}{cccc} \pi : & R^m & \longrightarrow & M \\ & e_i^m & \longmapsto & z_i. \end{array}$$

From the first isomorphism theorem we have that $M = \operatorname{Im} \pi \cong \dfrac{R^m}{\operatorname{Ker} \pi}$. To study M is in some sense to study $\operatorname{Ker} \pi$.

Definition 4.3.6. Let $\phi \colon M \to N$ be an R-linear map. The **cokernel** of ϕ, denoted by $\operatorname{Coker} \phi$, is defined to be $N/\operatorname{Im} \phi$.

Let $A_{m \times n}$ be a matrix over R. Define $\operatorname{Coker} A$ to be the quotient space of R^m modulo the column space of A.

Lemma 4.3.7. *Let $\phi \colon R^n \to R^m$ be an R-linear map. Let A be* any *matrix representing ϕ. Then* $\operatorname{Coker} \phi \cong \operatorname{Coker} A$.

Proof. Let $A = (a_{ij})_{m \times n}$ be the matrix representing ϕ with respect to the bases $\{u_1, u_2, \ldots, u_n\}$ and $\{v_1, v_2, \ldots, v_m\}$. Then

$$\operatorname{Im} \phi = \left\langle \sum_{i=1}^{m} a_{ij} v_i : j = 1, 2, \ldots, n \right\rangle.$$

There is a natural isomorphism $f \colon R^m \to R^m$ sending e_i^m to v_i for each i. It is easy to see the kernel of the epimorphism

$$R^m \xrightarrow{f} R^m \twoheadrightarrow R^m/\operatorname{Im} \phi$$

is the column space of A. Hence

$$\operatorname{Coker} A \cong R^m/\operatorname{Im} \phi = \operatorname{Coker} \phi$$

by the first isomorphism theorem. $\qquad\qquad\square$

Let's continue with our discussion on the finitely generated module M. Suppose $\operatorname{Ker} \pi$ as in (4.3.2) is finitely generated over R. We may find a finite set of generators $f_1, f_2, \ldots, f_n$ for $\operatorname{Ker} \pi$. Define the R-linear map

$$\begin{aligned} \phi \colon \quad R^n &\longrightarrow R^m \\ e_j^n &\longmapsto f_j. \end{aligned}$$

Then $\operatorname{Im} \phi = \operatorname{Ker} \pi$. It follows that $M \cong R^m/\operatorname{Im} \phi$ is the cokernel of ϕ. With respect to the ordered standard bases, ϕ is represented by the matrix $A = (f_1^{\mathbf{col}}, \ldots, f_n^{\mathbf{col}})_{m \times n}$ in column notation. We may thus view M as the cokernel of the matrix A. This fact can be depicted by the following diagram:

$$R^n \xrightarrow{\;A\;} R^m \twoheadrightarrow \operatorname{Coker} A.$$

Example 4.3.8. Let $M = \mathbb{Z} \oplus \mathbb{Z}/\langle(6,9), (2,2)\rangle$ as in Example 4.3.5. Let $\pi\colon \mathbb{Z}^2 \to M$ be the map sending $(1,0)$ to $\overline{(1,0)}$ and $(0,1)$ to $\overline{(0,1)}$. Then π is onto and the kernel of π is $\mathbb{Z}(6,9) + \mathbb{Z}(2,2)$. Let ϕ be the $\mathbb{Z}$-linear map

$$\begin{aligned} \phi\colon \quad \mathbb{Z}^2 \quad &\longrightarrow \quad \mathbb{Z}^2 \\ (1,0) \quad &\longmapsto \quad (6,9) \\ (0,1) \quad &\longmapsto \quad (2,2). \end{aligned}$$

Then $\operatorname{Im}\phi = \operatorname{Ker}\pi$. The matrix representing ϕ with respect to the ordered standard bases is

$$A = \begin{pmatrix} 6 & 2 \\ 9 & 2 \end{pmatrix}.$$

We have that $M \cong \operatorname{Coker} A$.

Lemma 4.3.9. *Let $A \in M_{m \times n}(R)$ and let P and Q be invertible matrices of size m and n respectively. Then $\operatorname{Coker} A \cong \operatorname{Coker} PAQ$.*

Proof. Since P and Q can be regarded as base change matrices, the matrix PAQ also represents ϕ with respect to a (possibly) different pair of bases. Now the result follows from Lemma 4.3.7. $\qquad\square$

What P does is to change how we describe elements in R^m by using a different base. The same thing can be said about Q. However, we would like to offer a different perspective on Q. We can think of AQ as performing column operations on A. Hence the columns of AQ are simply R-linear combinations of columns of A. Any column of AQ is an element in the column space of A. We conclude that the column space of AQ is a submodule of the column space of A. Similarly, the column space of $A = (AQ)Q^{-1}$ is also a submodule of that of AQ. Thus the columns of AQ are simply a different set of generators for the column space of A.

The problem now comes down to finding P and Q such that PAQ is as simple as possible. In other words, if we can find a "good" base for R^m and a "good" set of generators for the column space of A, we might be able to understand $\operatorname{Coker} A$ completely.

Finitely generated modules over a PID

The situation becomes noticeably more complicated if some submodule of the free module R^m is not finitely generated. In such a case we won't have any matrix at our disposal. Fortunately, this does not happen when R is a PID.

Proposition 4.3.10. *Let D be a PID. Then any submodule of D^n is free of rank $\leq n$.*

In order not to interrupt the flow of our discussion, we leave the proof of this proposition as an exercise. See Exercise 2.

For the rest of this section, we will assume that D is a PID and M is generated by m elements over D. Then $M \cong D^m / \operatorname{Ker} \pi$ for some R-linear map π. By Proposition 4.3.10, $\operatorname{Ker} \pi$ is free of finite rank. In particular, $\operatorname{Ker} \pi$ is finitely generated. Find a finite set of generators $f_1, f_2, \ldots, f_n$ for $\operatorname{Ker} \pi$. (This does not have to be a basis for $\operatorname{Ker} \pi$.) Let $\phi \colon D^n \to D^m$ be the D-linear map which maps the ordered standard basis onto $f_1, f_2, \ldots, f_n$. Thus $\operatorname{Im} \phi = \operatorname{Ker} \pi$ and $M \cong \operatorname{Coker} \phi \cong \operatorname{Coker} A$ where

$$A = (f_1^{\mathbf{col}}, \ldots, f_n^{\mathbf{col}})_{m \times n}.$$

By Theorem 4.2.6, we can find invertible matrices P and Q such that

$$PAQ = \operatorname{diag}\{d_1, d_2, \ldots, d_r, 0, \ldots, 0\}$$

where $d_1, d_2, \ldots, d_r$ are nonzero and $d_i | d_{i+1}$ for $i = 1, 2, \ldots, r - 1$. This means that we can find a basis $\{u_1, u_2, \ldots, u_m\}$ for D^m so that $\operatorname{Im} \phi$ is generated by $\{d_1 u_1, d_2 u_2, \ldots, d_r u_r\}$.

Lemma 4.3.11. *Let D be an integral domain and let $\{u_1, u_2, \ldots, u_m\}$ be a basis for D^m over D. Then $d_1 u_1, d_2 u_2, \ldots, d_r u_r$ are linearly independent over D when the d_i's are nonzero elements in D.*

Proof. Let $a_1, a_2, \ldots, a_r \in D$ be such that

$$a_1 d_1 u_1 + a_2 d_2 u_2 + \cdots + a_r d_r u_r = 0.$$

By assumption, $u_1, u_2, \ldots, u_m$ are linearly independent over D. So

$$a_1 d_1 = a_2 d_2 = \cdots = a_r d_r = 0.$$

Since D is an integral domain, we have that $a_1 = a_2 = \cdots = a_r = 0$. We conclude that $d_1 u_1, d_2 u_2, \ldots, d_r u_r$ are linearly independent over D.   □

Let's continue with our discussion of $D^m / \operatorname{Ker} \pi$. We have that D^m is the (internal) direct sum $Du_1 \oplus Du_2 \oplus \cdots \oplus Du_m$ since the u_i's form a basis for D^m over D. From Lemma 4.3.11 we have

$$\operatorname{Im} \phi = Dd_1 u_1 \oplus Dd_2 u_2 \oplus \cdots \oplus Dd_r u_r \oplus 0 \oplus \cdots \oplus 0.$$

Lemma 4.3.12. *Let D be a domain. Let z be a nonzero element in D^m and $d \in D$. Then*

$$\frac{Dz}{Ddz} \cong \frac{D}{Dd}.$$

In particular, $\operatorname{ann}_D \overline{z} = \operatorname{ann}_D \overline{1} = Dd$.

Proof. Let f be the composite of the following two D-linear maps.

$$
\begin{array}{ccccc}
D & \longrightarrow & Dz & \longrightarrow & Dz/Ddz \\
a & \longmapsto & az & \longmapsto & \overline{az}.
\end{array}
$$

The map f is surjective because it is the composite of two surjective maps. We claim that $\operatorname{Ker} f = Dd$.

Since $f(d) = \overline{dz} = 0$, we have $Dd \subseteq \operatorname{Ker} f$. On the other hand, let $a \in \operatorname{Ker} f$. Assume $z = (a_1, a_2, \ldots, a_m)$ where $a_k \neq 0$ for some k. Then

$$f(a) = \overline{az} = 0 \implies az \in Ddz \implies az = bdz \text{ for some } b \in D$$
$$\implies (a - bd)z = 0 \implies (a - bd)a_k = 0 \implies a - bd = 0 \implies a = bd \in Dd.$$

Thus $\operatorname{Ker} f = Dd$. We have $D/Dd \cong Dz/Ddz$ by the first isomorphism theorem.

Finally, notice that the element $\overline{z}$ in Dz/Ddz corresponds to the element $\overline{1}$ in D/Dd. They should share the same annihilator in D. It is easy to see that $\operatorname{ann}_D \overline{1} = Dd$. $\qquad \square$

Lemma 4.3.13. *Let R be a ring and let $M_1, M_2, \ldots, M_n$ be R-modules and let N_i be a submodule of M_i for each i. Then*

$$\frac{M_1 \oplus \cdots \oplus M_n}{N_1 \oplus \cdots \oplus N_n} \cong \frac{M_1}{N_1} \oplus \cdots \oplus \frac{M_n}{N_n}.$$

Proof. We leave it as an easy exercise to check that the following map

$$
\begin{array}{ccc}
f : \quad M_1 \oplus \cdots \oplus M_n & \longrightarrow & \dfrac{M_1}{N_1} \oplus \cdots \oplus \dfrac{M_n}{N_n} \\
(m_1, \ldots, m_n) & \longmapsto & (\overline{m}_1, \ldots, \overline{m}_n),
\end{array}
$$

is an R-linear map. It is also easy to see that f is onto. It remains to find $\operatorname{Ker} f$. We have that

$$f(m_1, \ldots, m_n) = (\bar{0}, \ldots, \bar{0}) \iff \overline{m}_i = \bar{0} \text{ for all } i \iff m_i \in N_i \text{ for all } i.$$

Hence $\operatorname{Ker} f = N_1 \oplus \cdots \oplus N_n$. The result follows from the first isomorphism theorem. $\qquad\qquad\square$

Using Lemma 4.3.13, we have

$$
\begin{aligned}
M &\cong \frac{D^m}{\operatorname{Im} \phi} = \frac{Du_1 \oplus \cdots \oplus Du_r \oplus Du_{r+1} \oplus \cdots \oplus Du_m}{Dd_1 u_1 \oplus \cdots \oplus Dd_r u_r \oplus 0 \oplus \cdots \oplus 0} \\
&\cong \frac{Du_1}{Dd_1 u_1} \oplus \cdots \oplus \frac{Du_r}{Dd_r u_r} \oplus Du_{r+1} \oplus \cdots \oplus Du_m.
\end{aligned}
$$

Let $K = Dd_1 u_1 \oplus \cdots \oplus Dd_r u_r \oplus 0 \oplus \cdots \oplus 0$. We will write $\overline{u}_i$ for $u_i + K$ in M. Next we try to find $\operatorname{ann}_D \overline{u}_i$ for each i.

For $1 \leq i \leq r$, clearly $d_i \in \operatorname{ann}_D \overline{u}_i$. Let $a \in D$ be such that $a\overline{u}_i = 0$. Then $au_i \in K$. We may find $b_1, b_2, \ldots, b_r \in D$ such that

$$au_i = b_1 d_1 u_1 + \cdots + b_r d_r u_r.$$

Since the u_i's form a basis for D^m over D, we have that $a = b_i d_i \in Dd_i$. We have shown that

$$\operatorname{ann}_D \overline{u}_i = Dd_i \qquad \text{for } 1 \leq i \leq r.$$

For $r + 1 \leq i \leq m$, if $a\overline{u}_i = 0$, we may find $b_1, b_2, \ldots, b_r \in D$ such that

$$au_i = b_1 d_1 u_1 + \cdots + b_r d_r u_r.$$

Since the u_i's form a basis for D^m over D, we have that $a = 0$. Hence

$$\operatorname{ann}_D \overline{u}_i = (0) \qquad \text{for } r + 1 \leq i \leq m.$$

Using Lemma 4.3.12 we have

$$
\begin{aligned}
M &\cong \frac{D}{Dd_1} \oplus \cdots \oplus \frac{D}{Dd_r} \oplus D \oplus \cdots \oplus D \\
&\cong \frac{D}{\operatorname{ann} \overline{u}_1} \oplus \cdots \oplus \frac{D}{\operatorname{ann} \overline{u}_r} \oplus \frac{D}{\operatorname{ann} \overline{u}_{r+1}} \oplus \cdots \oplus \frac{D}{\operatorname{ann} \overline{u}_m}.
\end{aligned}
$$

If we let $d_i = 0$ for $r + 1 \leq i \leq m$, we have that $\operatorname{ann}_D \overline{u}_i = (d_i)$. Remember that from Theorem 4.2.6, we have that

$$Dd_1 \supseteq Dd_2 \supseteq \cdots \supseteq Dd_r \supsetneq Dd_{r+1} = \cdots = Dd_m.$$

$$\parallel \qquad\qquad\qquad\qquad\qquad \parallel$$
$$(0) \qquad\qquad\qquad\qquad\qquad (0)$$

Suppose d_i are units for $1 \leq i \leq t$ while d_{t+1} is not a unit in D. Then $\overline{u}_i = 0$ for $1 \leq i \leq t$ while $\overline{u}_i \neq 0$ for $t + 1 \leq i \leq m$. Let $z_i = \overline{u}_{t+i}$ for $i = 1, \ldots, m - t$. We now have the main theorem of this section.

Theorem 4.3.14 (The structure theorem of finitely generated modules over a PID). *Let D be a PID and let M be a finitely generated nontrivial module over D. There exist $z_1, z_2, \ldots, z_s \in M$ such that*

$$M = Dz_1 \oplus Dz_2 \oplus \cdots \oplus Dz_s \cong \frac{D}{\operatorname{ann} z_1} \oplus \frac{D}{\operatorname{ann} z_2} \oplus \cdots \oplus \frac{D}{\operatorname{ann} z_s}$$

is a direct sum of cyclic modules in which

$$D \supsetneq \operatorname{ann} z_1 \supseteq \operatorname{ann} z_2 \supseteq \cdots \supseteq \operatorname{ann} z_s.$$

This theorem tells us that the best way to understand a finitely generated module over a PID is to view it as a direct sum of finitely many cyclic modules.

Example 4.3.15. Let $M = \mathbb{Z} \oplus \mathbb{Z}/\langle (6, 9), (2, 2) \rangle$ as in Example 4.3.5. Express M as a direct sum of cyclic modules as in Theorem 4.3.14. How many elements are there in M? Use these new results to find $\operatorname{ann}_{\mathbb{Z}}(4, 6)$ again.

Solution. The module M is the cokernel of the $\mathbb{Z}$-linear map

$$\begin{array}{rccc} \phi: & \mathbb{Z}^2 & \longrightarrow & \mathbb{Z}_2 \\ & (1, 0) & \longmapsto & (6, 9) \ , \\ & (0, 1) & \longmapsto & (2, 2) \end{array}$$

or we may say that M is the cokernel of

$$A = \begin{pmatrix} 6 & 2 \\ 9 & 2 \end{pmatrix} \in M_2(\mathbb{Z}).$$

First, we obtain the normal form for A. The g.c.d. of the entries in A is an associate to 1. We will try to produce 1 as an entry:

$$A_1 = A \begin{pmatrix} 1 & 1 \\ -4 & 1 \end{pmatrix} = \begin{pmatrix} -2 & 2 \\ 1 & 2 \end{pmatrix}.$$

Next move the entry 1 to the $(1,1)$-place:

$$A_2 = \begin{pmatrix} 0 & 1 \\ 1 & 0 \end{pmatrix} A_1 = \begin{pmatrix} 1 & 2 \\ -2 & 2 \end{pmatrix}.$$

Use the $(1,1)$-entry to cancel other entries in the first column and in the first row:

$$B = \begin{pmatrix} 1 & 0 \\ 2 & 1 \end{pmatrix} A_2 \begin{pmatrix} 1 & -2 \\ 0 & 1 \end{pmatrix} = \begin{pmatrix} 1 & 0 \\ 0 & 6 \end{pmatrix}.$$

We have obtained a normal form for A. Hence we have that $B = PAQ$ where

$$P = \begin{pmatrix} 1 & 0 \\ 2 & 1 \end{pmatrix} \begin{pmatrix} 0 & 1 \\ 1 & 0 \end{pmatrix}.$$

From the discussion after Lemma 4.3.9, we can see that the columns of AQ gives us a new (and more convenient) set of generators for $\mathrm{Im}\,\phi$, while P may be viewed as a base change of the codomain of ϕ from the standard ordered basis to a certain ordered basis (u_1, u_2). Since

$$P^{-1} = \begin{pmatrix} 0 & 1 \\ 1 & 0 \end{pmatrix} \begin{pmatrix} 1 & 0 \\ -2 & 1 \end{pmatrix} = \begin{pmatrix} -2 & 1 \\ 1 & 0 \end{pmatrix},$$

we have $u_1 = (-2, 1)$ and $u_2 = (1, 0)$. Moreover, $\mathrm{Im}\,\phi$ is generated by u_1 and $6u_2$. Modulo $\mathrm{Im}\,\phi$,

$$\mathrm{ann}_{\mathbb{Z}}\,\overline{(-2, 1)} = (1) \quad \text{and} \quad \mathrm{ann}_{\mathbb{Z}}\,\overline{(1, 0)} = (6).$$

Hence $\overline{(-2, 1)} = 0$ and $\overline{(1, 0)} \neq 0$ in M. We have that

$$M = \mathbb{Z}\overline{(1, 0)} \cong \frac{\mathbb{Z}}{(6)} \cong \mathbb{Z}_6, \quad \text{where } \mathrm{ann}\,\overline{(1, 0)} = (6).$$

This implies that there are exactly 6 elements in M. Since

$$\overline{(4, 6)} = \overline{6u_1 + 16u_2} = 16\overline{u_2} = -2\overline{u_2},$$

we conclude that $\mathrm{ann}_{\mathbb{Z}}\,\overline{(4, 6)} = (3)$. *Cf.* Example 4.3.5.   ◇

Example 4.3.16. Consider $G = \operatorname{Coker} A$ as an additive group where

$$A = \begin{pmatrix} 0 & 1 & 2 \\ 1 & 2 & 3 \\ 2 & 3 & 4 \\ 3 & 4 & 5 \end{pmatrix} \in M_{4\times 3}(\mathbb{Z}).$$

Find the order of G? How many elements are of finite order in G?

Solution. We may view G as a $\mathbb{Z}$-module. We will use the structure theorem to solve this problem. Normalize A as follows:

$$\begin{pmatrix} 0 & 1 & 2 \\ 1 & 2 & 3 \\ 2 & 3 & 4 \\ 3 & 4 & 5 \end{pmatrix} \rightsquigarrow \begin{pmatrix} 0 & 1 & 0 \\ 1 & 2 & -1 \\ 2 & 3 & -2 \\ 3 & 4 & -3 \end{pmatrix} \rightsquigarrow \begin{pmatrix} 0 & 1 & 0 \\ 1 & 2 & 0 \\ 2 & 3 & 0 \\ 3 & 4 & 0 \end{pmatrix}$$

$$\rightsquigarrow \begin{pmatrix} 1 & 0 & 0 \\ 2 & 1 & 0 \\ 3 & 2 & 0 \\ 4 & 3 & 0 \end{pmatrix} \rightsquigarrow \begin{pmatrix} 1 & 0 & 0 \\ 0 & 1 & 0 \\ 0 & 2 & 0 \\ 0 & 3 & 0 \end{pmatrix} \rightsquigarrow \begin{pmatrix} 1 & 0 & 0 \\ 0 & 1 & 0 \\ 0 & 0 & 0 \\ 0 & 0 & 0 \end{pmatrix}.$$

Thus

$$G = \operatorname{Coker} A \cong \frac{\mathbb{Z}}{(1)} \oplus \frac{\mathbb{Z}}{(1)} \oplus \frac{\mathbb{Z}}{(0)} \oplus \frac{\mathbb{Z}}{(0)} \cong \mathbb{Z}^2.$$

We conclude that G is an infinite group and there is exactly one element of finite order in G, the trivial element. ◇

Example 4.3.17. In this example we will try to determine the structure of $M = \mathbb{Z}_{12}^2/K$ where K is generated by $f_1 = (\overline{3}, \overline{3})$, $f_2 = (\overline{6}, \overline{2})$ and $f_3 = (\overline{0}, \overline{4})$.

Even though $\mathbb{Z}_{12}$ is not a domain, it is still a *principal ideal ring* since its ideals are all principal. The $\mathbb{Z}_{12}$-module $M = \operatorname{Coker} A$ where

$$A = \begin{pmatrix} \overline{3} & \overline{6} & \overline{0} \\ \overline{3} & \overline{2} & \overline{4} \end{pmatrix}.$$

From our previous discussion, applying elementary operations on A does not alter the structure of $\operatorname{Coker} A$. Let's see if we can normalize A even though $\mathbb{Z}_{12}$ is not a PID:

$$A \rightsquigarrow \begin{pmatrix} -\overline{3} & \overline{6} & \overline{0} \\ \overline{1} & \overline{2} & \overline{4} \end{pmatrix} \rightsquigarrow \begin{pmatrix} \overline{0} & \overline{0} & \overline{0} \\ \overline{1} & \overline{0} & \overline{0} \end{pmatrix} \rightsquigarrow \begin{pmatrix} \overline{1} & \overline{0} & \overline{0} \\ \overline{0} & \overline{0} & \overline{0} \end{pmatrix}.$$

Thus

$$M = \operatorname{Coker} A \cong \frac{\mathbb{Z}_{12}}{\mathbb{Z}_{12}} \oplus \frac{\mathbb{Z}_{12}}{(0)} \cong \mathbb{Z}_{12}$$

contains exactly 12 elements.

To understand M, we may also view it as a $\mathbb{Z}$-module. By Lemma 4.3.13,

$$\mathbb{Z}_{12} \oplus \mathbb{Z}_{12} \cong \frac{\mathbb{Z} \oplus \mathbb{Z}}{12\mathbb{Z} \oplus 12\mathbb{Z}} \cong \frac{\mathbb{Z} \oplus \mathbb{Z}}{\langle (12, 0), (0, 12) \rangle}.$$

By the third isomorphism theorem we have that

$$M \cong \frac{\mathbb{Z} \oplus \mathbb{Z}}{\langle (12, 0), (0, 12), (3, 3), (6, 2), (0, 4) \rangle}$$

is the cokernel of

$$B = \begin{pmatrix} 3 & 6 & 0 & 12 & 0 \\ 3 & 2 & 4 & 0 & 12 \end{pmatrix},$$

over $\mathbb{Z}$. We have that

$$\begin{pmatrix} 1 & 0 & 0 & 0 & 0 \\ 0 & 12 & 0 & 0 & 0 \end{pmatrix},$$

is a normal form of B. Hence

$$M \cong \operatorname{Coker} B \cong \frac{\mathbb{Z}}{\mathbb{Z}} \oplus \frac{\mathbb{Z}}{12\mathbb{Z}} \cong \frac{\mathbb{Z}}{12\mathbb{Z}} \cong \mathbb{Z}_{12}$$

as a $\mathbb{Z}$-module.

Exercises 4.3

1. Let M be a module over the ring R and N be an R-submodule of M. Let $x \in M$.

   (a) Show that $\operatorname{ann} x \subseteq \operatorname{ann}(x + N)$.

   (b) Show that $\operatorname{ann} x = \operatorname{ann}(x + N)$ if and only if $Rx \cap N = \langle 0 \rangle$.

2. Prove Proposition 4.3.10 in the following steps.

   (a) First, verify this proposition for the case $n = 1$. Remember that D is a PID.

(b) Let N be a submodule of D^{n+1} and let

$$N' = N \cap \big((0) \times D^n\big).$$

First check that N' is a submodule of $(0) \times D^n \cong D^n$ and conclude that N' is free with rank $\leq n$ using the induction hypothesis. Find a basis $\{f_1, f_2, \ldots, f_s\}$, $s \leq n$, for N' over D.

(c) Let $I = \{a \in D : a$ is the first coordinate of some element in $N\}$. Show that I is an ideal of R. Since D is a PID, we can find $d \in D$ such that $I = (d)$.

(d) If $d = 0$, then $N = N'$. Conclude that N is free of rank $\leq n$.

(e) If $d \neq 0$, find $f \in N$ such that d is the first coordinate of f. Show that $f, f_1, f_2, \ldots, f_s$ form a basis for N over D. Conclude that N is free of rank $\leq n+1$.

3. Determine the structure of $\mathbb{Z}^3/K$ over $\mathbb{Z}$ where K is generated by $f_1 = (2, 1, -3)$ and $f_2 = (1, -1, 2)$.

4. Let $D = \mathbb{Z}[i]$. Determine the structure of $M = D^3/K$ over D where K is generated by $f_1 = (1, 3, 6)$, $f_2 = (2 + 3i, -3i, 12 - 18i)$ and $f_3 = (2 - 3i, 6 + 9i, -18i)$. Find the order of M.

5. Consider $\mathbb{Z}[x]$, the polynomial ring of one variable over $\mathbb{Z}$. Let I be the ideal generated by 2 and x in $\mathbb{Z}[x]$. Show that I is not a principal ideal and show that I is not a direct sum of cyclic modules over $\mathbb{Z}[x]$.

4.4   Invariance theorem

In this section, we will concentrate on the "uniqueness" part of the structure theorem of finitely generated modules over a PID (Theorem 4.3.14). See Theorem 4.4.21.

The torsion submodule and torsion free rank

Definition 4.4.1. Let R be a ring and M be an R-module. We say an element m in an R-module is **torsion** if $\mathrm{ann}_R m \neq (0)$. Define $\mathrm{tor}_R M$, or simply $\mathrm{tor}\, M$ if R is understood, to be the set $\{m \in M : \mathrm{ann}_R m \neq (0)\}$.

In short, a torsion element is an element which can be "killed" by a nonzero element in R.

Lemma 4.4.2. *Let D be a domain and M be a D-module. Then $\operatorname{tor}_D M$ is a submodule of M over D.*

Proof. Let $d \in D$ and let m, $n \in \operatorname{tor} M$. Find $a \in \operatorname{ann} m \setminus (0)$. Then $a(dm) = d(am) = 0$. We have $dm \in \operatorname{tor} M$. Find $b \in \operatorname{ann} n \setminus (0)$ as well. Note that $ab(m + n) = b(am) + a(bn) = 0$. We have $ab \neq 0$ since D is a domain. Hence $m + n \in \operatorname{tor} M$. We have shown that $\operatorname{tor} M$ is a submodule of M over D. $\qquad\square$

Definition 4.4.3. Let M be a module over a domain D. The submodule $\operatorname{tor}_D M$ is called the **torsion submodule** of M over D. We say that M is a **torsion module** over D if $M = \operatorname{tor}_D M$.

Example 4.4.4. As a $\mathbb{Z}$-modules, $\mathbb{Z}_2$ is a torsion module. However, as a $\mathbb{Z}_2$-module, $\mathbb{Z}_2$ is not torsion since $\operatorname{ann}_{\mathbb{Z}_2} \overline{1} = (0)$.

Let D be a PID and let M be a finitely generated module over D. Using Theorem 4.3.14, we may find $z_1, \ldots, z_s$ in M such that

$$(4.4.1) \qquad M = Dz_1 \oplus Dz_2 \oplus \cdots \oplus Dz_s$$

and the order ideals satisfies

$$D \supsetneq \operatorname{ann} z_1 \supseteq \operatorname{ann} z_2 \supseteq \cdots \supseteq \operatorname{ann} z_r \supsetneq \operatorname{ann} z_{r+1} = \cdots = \operatorname{ann} z_s = (0).$$

We leave it as an easy exercises to show that

$$\operatorname{tor} M = Dz_1 \oplus Dz_2 \oplus \cdots \oplus Dz_r$$

(see Exercise 1). It is clear that $Dz_{r+1} \oplus \cdots \oplus Dz_s$ is a free module of rank $s - r$. We thus have the following result.

Lemma 4.4.5. *Any finitely generated module over a PID is a direct sum of its torsion submodule and a free submodule.*

The torsion submodule does not depend on the decomposition. On the other hand, the choice of direct summand in Lemma 4.4.5 which is a free module is not unique. It is possible to find distinct free submodules F_1 and

F_2 of M such that $M = \operatorname{tor} M \oplus F_1 = \operatorname{tor} M \oplus F_2$. However, the ranks of F_1 and of F_2 must be the same because both summands are isomorphic to $M/\operatorname{tor} M$, and the rank of a free module of finite rank is well-defined by Proposition 3.4.11.

Definition 4.4.6. Let M be a finitely generated module over a PID. We call the free rank of $M/\operatorname{tor} M$ the **torsion free rank** of M.

Example 4.4.7. Consider $M = \mathbb{Z}_6 \oplus \mathbb{Z}_8 \oplus \mathbb{Z}$ as a $\mathbb{Z}$-module. The torsion submodule $\operatorname{tor}_\mathbb{Z} M = \mathbb{Z}_6 \oplus \mathbb{Z}_8 \oplus (0) = \mathbb{Z}(\overline{1}, 0, 0) \oplus \mathbb{Z}(0, \overline{1}, 0)$. It is clear that $M = (\operatorname{tor}_\mathbb{Z} M) \oplus \mathbb{Z}(0, 0, 1)$ and $\mathbb{Z}(0, 0, 1) \cong \mathbb{Z}$ is a free $\mathbb{Z}$-module. We claim that $M = (\operatorname{tor}_\mathbb{Z} M) \oplus \mathbb{Z}(0, \overline{1}, 1)$ is a different decomposition of M into a direct sum of its torsion submodule and a free module over $\mathbb{Z}$. This shows that the free summand in a decomposition of M is not unique.

Note that for any $\ell, m, n \in Z$, we have that

$$(\overline{\ell}, \overline{m}, n) = \ell(\overline{1}, 0, 0) + (m - n)(\overline{0}, \overline{1}, 0) + n(0, \overline{1}, 1).$$

This shows that $M = \mathbb{Z}(\overline{1}, 0, 0) + \mathbb{Z}(0, \overline{1}, 0) + \mathbb{Z}(0, \overline{1}, 1)$. Suppose

$$(\overline{\ell}, 0, 0) + (0, \overline{m}, 0) + (0, \overline{n}, n) = (\overline{\ell}, \overline{m + n}, n) = (0, 0, 0).$$

This implies that $n = 0$, $\overline{\ell} = 0$ and $\overline{m} = \overline{m + n} = 0$. We have just shown that $M = \mathbb{Z}(\overline{1}, 0, 0) \oplus \mathbb{Z}(0, \overline{1}, 0) \oplus \mathbb{Z}(0, \overline{1}, 1)$.

Finally, let $k \in \mathbb{Z}$ be such that $k(0, \overline{1}, 1) = (0, \overline{k}, k) = (0, 0, 0)$. Observing the third coordinate we have that $k = 0$. Hence $\operatorname{ann}_\mathbb{Z}(0, \overline{1}, 1) = (0)$. We conclude that $\mathbb{Z}(0, \overline{1}, 1) \cong \mathbb{Z}$ is a free $\mathbb{Z}$-module. Our claim is thus established. It is possible to find other free summands for M. See Exercise 2.

Incidentally, the torsion free rank of M is 1.

From the discussion above we can make the following conclusion.

Proposition 4.4.8. *A finitely generated module over a PID is the direct sum of a torsion module and a free submodule. The torsion module is its torsion submodule. The rank of the free submodule is its torsion free rank. The torsion submodule and the torsion free rank are uniquely determined by the given module.*

The p-component and primary decomposition

Next we will concentrate on determining the "uniqueness" of decomposition of the torsion submodule. For this purpose, we will assume that the module in question is torsion for the time being.

Before we actually start, we need the more general version of *Chinese remainder theorem* (CRT for short).

Proposition 4.4.9 (Chinese remainder theorem). *Let D be a PID and let g, h be nonzero elements in D such that $(g, h) \sim 1$.*

(a) *Let $M = Dx$ where $\operatorname{ann} x = (gh)$. Then $M = Dy \oplus Dz$ for some y and $z \in M$ such that $\operatorname{ann} y = (g)$ and $\operatorname{ann} z = (h)$.*

(b) *Let $M = Dy + Dz$ where $\operatorname{ann} y = (g)$ and $\operatorname{ann} z = (h)$. Then there exists $x \in M$ such that $\operatorname{ann} x = (gh)$ and $M = Dy \oplus Dz = Dx$.*

In either case, we have that

$$\frac{D}{(gh)} \cong \frac{D}{(g)} \oplus \frac{D}{(h)}.$$

Proof. Since D is a PID, we may find a, $b \in D$ such that $ag + bh = 1$.

(a) Let $y = hx$ and $z = gx$. We first find $\operatorname{ann} y$ and $\operatorname{ann} z$.

Remember that $\operatorname{ann} x = (gh)$. Hence $gy = ghx = 0$. This implies that $(g) \subseteq \operatorname{ann} y$. Suppose $cy = 0$ for some $c \in D$. Then

$$chx = 0 \implies ch \in (gh) \implies ch = dgh \text{ for some } d \in D \implies c = dg \in (g)$$

since $h \neq 0$ and D is a domain. This says that $\operatorname{ann} y \subseteq (g)$. We conclude that $\operatorname{ann} y = (g)$. The argument for $\operatorname{ann} z = (h)$ is similar.

Next, we show that $Dx = Dy \oplus Dz$. Clearly, $Dy + Dz \subseteq Dx$. Conversely, we have $x = (ag + bh)x = by + az$ and so $Dx \subseteq Dy + Dz$. We conclude that $Dx = Dy + Dz$.

Suppose $cy + dz = 0$ for some c and $d \in D$. Then

$$(4.4.2) \qquad cy = (ag + bh)cy = bhcy = bhcy + bhdz = bh(cy + dz) = 0$$

from the facts $\operatorname{ann} y = (g)$ and $\operatorname{ann} z = (h)$. Similarly we also have $dz = 0$. Thus, Dy and Dz are independent over D and $Dx = Dy \oplus Dz$.

(b) Let $cy + dz = 0$ for some c and $d \in D$. The same argument as in
(4.4.2) shows that $cy = dz = 0$. This says that Dy and Dz are independent
over D and so $Dy + Dz = Dy \oplus Dz$.

Let $x = by + az$. Obviously $Dx \subseteq Dy + Dz$. We also have

$$hx = h(by + az) = bhy = bhy + agy = (ag + bh)y = y$$

since $gy = hz = 0$. Similarly, we have

$$gx = g(by + az) = agz = agz + bhz = (ag + bh)z = z.$$

This shows that y, $z \in Dx$ and $Dy + Dz \subseteq Dx$. We may conclude that
$Dx = Dy + Dz = Dy \oplus Dz$.

Finally, let's find $\operatorname{ann} x$. Since

$$ghx = gh(by + az) = hbgy + aghz = 0,$$

we have $(gh) \subseteq \operatorname{ann} x$. Conversely, let $cx = 0$. Then

$$0 = cx = cby + caz$$
$$\implies cby = caz = 0, \qquad \text{for } Dy \text{ and } Dz \text{ are independent over } D,$$
$$\implies cb \in (g) \text{ and } ca \in (h).$$

We thus have that $g|cb$ and $h|ca$ in D. Since $ag + bh = 1$, we have $(b, g) \sim 1$
and $(a, h) \sim 1$. It follows that $g|c$ and $h|c$ from Exercise 2(a), §4.2. Hence
$gh|c$ from Exercise 2(b), §4.2. This shows that $\operatorname{ann} x \subseteq (gh)$. We conclude
that $\operatorname{ann} x = (gh)$.

The last part of the statement follows from Lemma 4.3.2. $\qquad \square$

Lemma 4.4.10. *Let D be a PID. Let M be a D-module and let $x \in D$.
Suppose $\operatorname{ann} x = (d)$ where $d = u p_1^{e_1} p_2^{e_2} \cdots p_t^{e_t}$ such that u is a unit in
D, $e_i > 0$ for all i and the p_i's are distinct primes in D. Then there
exist $x_1, x_2, \ldots, x_t$ in D such that $Dx = Dx_1 \oplus Dx_2 \oplus \cdots \oplus Dx_t$ where
$\operatorname{ann} x_i = (p_i^{e_i})$.*

Proof. We will prove this lemma by induction on t. When $t = 1$, there is
nothing to prove. Assume $t \geq 2$. Since $u p_1^{e_1}$ and $p_2^{e_2} \cdots p_t^{e_t}$ are relatively

prime, by the Chinese remainder theorem, we may find x_1, $y \in Dx$ such that $Dx = Dx_1 \oplus Dy$ where

$$\text{ann } x_1 = (up_1^{e_1}) = (p_1^{e_1}) \text{ and } \text{ann } y = (p_2^{e_2} \cdots p_t^{e_t}).$$

The rest follows from the induction hypothesis.   □

Definition 4.4.11. Let D be a PID. We say a *cyclic* torsion module over D is **primary** if its order ideal is of the form (p^e) for some prime p in D. A decomposition into a direct sum of primary cyclic modules is called a **primary decomposition**.

Suppose given a finitely generated torsion module over a PID D. From the structure theorem (Theorem 4.3.14), we may find nonzero elements $z_1, z_2, \ldots, z_s$ in M such that

$$M = Dz_1 \oplus Dz_2 \oplus \cdots \oplus Dz_s$$

where $\text{ann } z_i = (d_i) \neq (0)$ for all i. Since each d_i can have only a finite number of prime factors, we may find finitely many distinct primes $p_1, p_2, \ldots, p_n$ in D such that

$$d_i = p_1^{e_{i1}} p_2^{e_{i2}} \cdots p_n^{e_{in}}$$

where $e_{ij} \geq 0$ for all i, j. From Lemma 4.4.10, we may find w_{ij} in M so that

$$(4.4.3) \qquad\qquad M = \bigoplus_i Dz_i = \bigoplus_{i,j} Dw_{ij}$$

where $\text{ann } w_{ij} = (p_j^{e_{ij}})$. Note that we allowed e_{ij} to be 0 for some (i, j) in which case $w_{ij} = 0$. We may drop these trivial terms without affecting the decomposition. We sum up the result in the following proposition.

Proposition 4.4.12. *Any finitely generated torsion module over a PID is a direct sum of primary cyclic modules. In other words, any finitely generated torsion module over a PID has a primary decomposition.*

One of our goals in this section is to explain why the primary decomposition of a finitely generated module over a PID is essentially unique. For this we will need the following concept.

Definition 4.4.13. Let D be a UFD, let M be module over D and let p be a prime in D. Define the p-**component** M_p of M to be the subset

$$\{m \in M : p^k m = 0 \text{ for some } k = 0, 1, 2, 3, \ldots\}.$$

Lemma 4.4.14. *Let D be a UFD, let M be a module over D and let p be a prime in D. The subset M_p is a submodule of M which does not depend on any decomposition of M.*

Proof. Suppose given $a \in D$ and $m, n \in M_p$. Find i high enough such that $p^i m = p^i n = 0$. Then

$$p^i(m + n) = p^i m + p^i n = 0, \quad \text{and} \quad p^i(am) = a(p^i m) = 0.$$

This shows that $m + n$ and am are both in M_p. Thus M_p is a submodule of M over D. □

Lemma 4.4.15. *Let D be a PID, let M be a module over D and let $m \in M$. Suppose c and d are relatively prime in D. If $cm = dm = 0$, then $m = 0$.*

Proof. Since D is a PID, we may find a, b in D such that $ac + bd = 1$. Thus $m = (ac + bd)m = acm + bdm = 0$. □

Lemma 4.4.16. *Let D be a PID, let M be a D-module and let p be a prime in D. An element $m \in M_p$ if and only if $\operatorname{ann}_D m = p^e D$ for some non-negative integer e.*

Proof. Without loss of generality we may assume $m \neq 0$.

The "if" part: This is true by definition of M_p.

The "only if" part: Let $m \in M_p$ where $\operatorname{ann} m = p^e k D$ for some non-negative integer e and $(k, p) \sim 1$. Suppose k is not a unit. Then $p^e \notin \operatorname{ann} m$. Thus $p^e m \neq 0$. Since $m \in M_p$, there exists an integer $f > e$ such that $p^f m = 0$. Note that $k(p^e m) = 0$ and $p^{f-e}(p^e m) = 0$. By Lemma 4.4.15 we have that $p^e m = 0$, a contradiction. Hence we conclude that k is a unit and $\operatorname{ann} m = p^e D$. □

Lemma 4.4.17. *Let D be a PID, let $p, \ldots, p_n$ be distinct primes in D and let M be a module over D. Then $M_{p_1}, \ldots, M_{p_n}$ are independent over D.*

Proof. Let $m_i \in M_{p_i}$ for each i such that $m_1 + m_2 + \cdots + m_n = 0$. Find N large enough so that $p_i^N m_i = 0$ for all i. Then

$$0 = p_1^N \cdots \widehat{p_i^N} \cdots p_n^N (m_1 + m_2 + \cdots + m_n) = p_1^N \cdots \widehat{p_i^N} \cdots p_n^N m_i$$

for each i. (The $\widehat{}$ above p_i^N denotes that the product is missing the term p_i^N.) Since p_i^N and $p_1^N \cdots \widehat{p_i^N} \cdots p_n^N$ are relatively prime in D, from Lemma 4.4.15 we have that $m_i = 0$. This is true for all i. Thus $M_{p_1}, \ldots, M_{p_n}$ are independent over D. $\qquad\square$

Lemma 4.4.18. *Let D be a PID and let M be a finitely generated torsion D-module. From Theorem 4.3.14, there exist $z_1, z_2, \ldots, z_s \in M$ such that*

$$M = Dz_1 \oplus Dz_2 \oplus \cdots \oplus Dz_s$$

where $\operatorname{ann} z_i = (d_i) \neq (0)$ *for all i and $d_i | d_{i+1}$ for $i = 1, 2, \ldots s - 1$. Let $m \in M$. If $\operatorname{ann}_D m = (a)$, then $a | d_s$ in D. In particular, we have $d_s m = 0$.*

Proof. Note that $d_s z_i = 0$ for all i since $d_s \in (d_i) = \operatorname{ann}_D z_i$. Hence $d_s m = 0$ for all $m \in M$ since m is a D-linear combination of the z_i's. Thus $(d_s) \subseteq (a)$, or equivalently, $a | d_s$. $\qquad\square$

Proposition 4.4.19. *Let D be a PID and let M be a finitely generated torsion module over D. We will use the setup in Lemma 4.4.18. Let $p_1, p_2, \ldots, p_n$ be all the distinct prime factors of d_s. The following statements are true.*

(a) *If p is any prime in D which is distinct from $p_1, p_2, \ldots, p_n$, then $M_p = 0$.*

(b) *The module $M = M_{p_1} \oplus M_{p_2} \oplus \cdots \oplus M_{p_n}$.*

Proof. (a) Let $m \in M_p$. By Lemma 4.4.18, we have $d_s m = 0$. Find a positive integer e such that $p^e m = 0$. From Lemma 4.4.15 we have that $m = 0$ since d_s and p^e are relatively prime with each other.

(b) Let $m \in M$. Let $\operatorname{ann} m = (a)$ for some $a \in D$. From Lemma 4.4.18, we have $a | d_s$. The prime factors of a are also prime factors of d_s. By Lemma 4.4.10, we may find $m_1, m_2, \ldots, m_n$ and positive integers e_i such that $m = m_1 + m_2 + \cdots + m_n$ where $p_i^{e_i} m_i = 0$ for each i. Of course, some of the m_i's may well be 0. Thus $m \in M_{p_1} + M_{p_2} + \cdots + M_{p_n}$.

We conclude that $M = M_{p_1} + M_{p_2} + \cdots + M_{p_n}$. The rest follows from Lemma 4.4.17. $\square$

Corollary 4.4.20. *Let D be a PID and let M be a finitely generated torsion module M over D. Suppose given a primary decomposition*

$$M = \bigoplus_{1 \leq i \leq n} \left(Dw_{i1} \oplus Dw_{i2} \oplus \cdots \oplus Dw_{it_i} \right)$$

where $p_1, p_2, \ldots, p_n$ are distinct primes in D and $\mathrm{ann}_D\, w_{ij} = (p_i^{e_{ij}})$ such that $e_{ij} > 0$ for all (i, j). Then

$$M_{p_i} = Dw_{i1} \oplus Dw_{i2} \oplus \cdots \oplus Dw_{it_i}$$

for all i. If p is a prime in D which is distinct from p_i for any i, then $M_p = 0$.

We leave this corollary as an easy exercise. See Exercise 3.

Let's review what we have achieved so far. When given a decomposition of a finitely generated module over a PID in the manner as in Theorem 4.3.14, we can extract two pieces of information unique to the given module: (i) the torsion submodule and (ii) the torsion free rank. We may use the Chinese remainder theorem to further break down the torsion submodule into a primary decomposition. From there we may collect the primary cyclic modules into p-components.

Invariance theorem

We are now ready for the main result of this section.

Theorem 4.4.21 (Invariance theorem). *Let D be a PID and M be a finitely generated torsion module over D such that*

$$M = Dz_1 \oplus Dz_2 \oplus \cdots \oplus Dz_s = Dw_1 \oplus Dw_2 \oplus \cdots \oplus Dw_t.$$

(a) *Suppose*

$$(1) \neq \mathrm{ann}\, z_1 \supseteq \mathrm{ann}\, z_2 \supseteq \cdots \supseteq \mathrm{ann}\, z_s \neq (1), \ \text{ and}$$
$$(0) \neq \mathrm{ann}\, w_1 \supseteq \mathrm{ann}\, w_2 \supseteq \cdots \supseteq \mathrm{ann}\, w_t \neq (0).$$

Then $s = t$ and $\mathrm{ann}\, z_i = \mathrm{ann}\, w_i$ for all i.

(b) *If the two decompositions are primary, then $s = t$ and after a rearrangement of terms, we have* $\operatorname{ann} z_i = \operatorname{ann} w_i$ *for all i.*

Definition 4.4.22. Let M be a finitely generated torsion module over a PID. We call the order ideals in Theorem 4.4.21(a) the **invariant factor ideals** of M. We call the order ideals in Theorem 4.4.21(b) the **elementary divisor ideals**. In general, the invariant factor ideals (or elementary divisor ideals, respectively) of a finitely generated M over a PID are defined to be the invariant factor ideals (or elementary divisor ideals) of its torsion submodule tor M.

In the case where $D = \mathbb{Z}$, we can choose a *unique* positive integer as the generator for a nontrivial ideal. In the case of a polynomial ring of one variable over a field, we can choose a *unique* monic polynomial as the generator for a nontrivial ideal. We will call these type of generators **normalized** generators. We will call the normalized generators (if possible) of the invariant factor ideas and the elementary divisor ideals the **invariant factors** and **elementary divisors** respectively.

Repeating invariant factors and elementary divisors appearing in the decompositions in Theorem 4.4.21 must be all present and accounted for.

Example 4.4.23. Let $M = \mathbb{Z}_{12} \oplus \mathbb{Z}_{20} \oplus \mathbb{Z}^2$ be a $\mathbb{Z}$-module. The torsion submodule tor $M = \mathbb{Z}_{12} \oplus \mathbb{Z}_{20} \oplus 0 \oplus 0$ and the torsion free rank of M is 2. Apply CRT to tor M to obtain

$$
\begin{aligned}
\operatorname{tor} M &= \mathbb{Z}(\overline{3}, \overline{0}, \overline{0}, \overline{0}) \oplus \mathbb{Z}(\overline{4}, \overline{0}, \overline{0}, \overline{0}) \oplus \mathbb{Z}(\overline{0}, \overline{5}, \overline{0}, \overline{0}) \oplus \mathbb{Z}(\overline{0}, \overline{4}, \overline{0}, \overline{0}) \\
&\cong \frac{\mathbb{Z}}{(4)} \oplus \frac{\mathbb{Z}}{(3)} \oplus \frac{\mathbb{Z}}{(4)} \oplus \frac{\mathbb{Z}}{(5)}.
\end{aligned}
$$

There are three p-components for tor M: $M_2 = \mathbb{Z}(\overline{3}, \overline{0}, \overline{0}, \overline{0}) \oplus \mathbb{Z}(\overline{0}, \overline{5}, \overline{0}, \overline{0})$, $M_3 = \mathbb{Z}(\overline{4}, \overline{0}, \overline{0}, \overline{0})$ and $M_5 = \mathbb{Z}(\overline{0}, \overline{4}, \overline{0}, \overline{0})$. The elementary divisors of tor M (or of M) are 4, 4, 3, 5. By CRT again, we have

$$
\operatorname{tor} M \cong \frac{\mathbb{Z}}{(4)} \oplus \left(\frac{\mathbb{Z}}{(3)} \oplus \frac{\mathbb{Z}}{(4)} \oplus \frac{\mathbb{Z}}{(5)} \right) \cong \frac{\mathbb{Z}}{(4)} \oplus \frac{\mathbb{Z}}{(60)}.
$$

The invariant factors of M are 4 and 60.

When one has a primary decomposition, one can use the CRT to recreate (in a unique way) a decomposition of the form as in Theorem 4.4.21(a).

Example 4.4.24. Find the invariant factors of

$$M = \mathbb{Z}_2 \oplus \mathbb{Z}_8 \oplus \mathbb{Z}_9 \oplus \mathbb{Z}_{27} \oplus \mathbb{Z}_{81} \oplus \mathbb{Z}_{125} \qquad \text{over } \mathbb{Z}.$$

Solution. We suggest that we make an array of the elementary divisors of M where the powers are arranged from low to high, flush right:

the row of 2:		2	8
the row of 3:	9	27	81 .
the row of 5:			125

By CRT we have that

$$M = \mathbb{Z}_9 \oplus \big(\mathbb{Z}_2 \oplus \mathbb{Z}_{27}\big) \oplus \big(\mathbb{Z}_8 \oplus \mathbb{Z}_{81} \oplus \mathbb{Z}_{125}\big).$$

The invariant factors of M are 9, 2·27 and 8·81·125. Reader please note that this is the only choice for the invariant factors to satisfy the requirement in Theorem 4.4.21(a).   ◇

Example 4.4.25. Let $D = \mathbb{R}[\lambda]$ be the polynomial ring of one variable over $\mathbb{R}$. Suppose the D-module M is a direct sum of cyclic modules whose order ideals are generated by the polynomials $(\lambda - 1)^3$, $(\lambda - 1)(\lambda^2 + 1)^4$, $(\lambda^2 + 1)^2$ and $(\lambda + 2)(\lambda^2 + 1)^2$. From the CRT we may obtain a set of elementary divisors for M as follows

$$(\lambda - 1)^3, \ (\lambda - 1), \ (\lambda^2 + 1)^4, \ (\lambda^2 + 1)^2, \ (\lambda + 2) \text{ and } (\lambda^2 + 1)^2.$$

Arrange the elementary divisors in the following manner:

$$\begin{array}{ccc} & \lambda - 1 & (\lambda - 1)^3 \\ (\lambda^2 + 1)^2 & (\lambda^2 + 1)^2 & (\lambda^2 + 1)^4 \\ & \lambda + 2 & \end{array} .$$

We may obtain the invariant factors of M as

$$(\lambda^2 + 1)^2, \ (\lambda - 1)(\lambda^2 + 1)^2 \text{ and } (\lambda - 1)^3(\lambda^2 + 1)^4(\lambda + 2).$$

Now change D to $\mathbb{C}[\lambda]$. The elementary divisors of M are

$$(\lambda - 1)^3, \ (\lambda - 1), \ (\lambda + i)^4, \ (\lambda - i)^4,$$

$$(\lambda + i)^2, \ (\lambda - i)^2, \ (\lambda + 2), \ (\lambda + i)^2 \text{ and } (\lambda - i)^2.$$

Arrange the elementary divisors in the following manner:

$$
\begin{array}{ccc}
& \lambda - 1 & (\lambda - 1)^3 \\
(\lambda + i)^2 & (\lambda + i)^2 & (\lambda + i)^4 \\
(\lambda - i)^2 & (\lambda - i)^2 & (\lambda - i)^4 \\
& \lambda + 2 &
\end{array}
.
$$

We can see that the invariant factors of M over $\mathbb{R}[\lambda]$ and over $\mathbb{C}[\lambda]$ are the same.

To prove the Invariance theorem, the last piece of the puzzles is the following lemma.

Lemma 4.4.26. *Let D be a PID and let M be a finitely generated D-module such that $M = M_p$ for some prime p of D. Suppose*

$$M = Dz_1 \oplus Dz_2 \oplus \cdots \oplus Dz_s = Dw_1 \oplus Dw_2 \oplus \cdots \oplus Dw_t$$

where $\operatorname{ann} z_i = (p^{e_i})$ *and* $\operatorname{ann} w_j = (p^{f_j})$. *Assume*

$$0 < e_1 \le e_2 \le \cdots \le e_s \quad and \quad 0 < f_1 \le f_2 \le \cdots \le f_t.$$

Then $s = t$ and $e_i = f_i$ for all i.

We leave the proof of this lemma as an exercise (see Exercise 10).

Proof of Theorem 4.4.21. (b) To obtain a primary decomposition, we first decompose M into a direct sum of p-components by Proposition 4.4.19. Note that the construction of the p-components depends only on the properties of elements in M, and not on any decomposition. We then further decompose each p-component into a direct sum of p-primary components by Corollary 4.4.20. The elementary divisor ideals in a primary decomposition of M are thus uniquely determined by Lemma 4.4.26.

(a) Suppose $M_{p_1}, M_{p_2}, \ldots, M_{p_n}$ are the only nontrivial p-components of M where $p_1, p_2, \ldots, p_n$ are distinct primes in D. Decompose each p-component

$$M_{p_i} = Dw_{i1} \oplus Dw_{i2} \oplus \cdots \oplus Dw_{it_i}$$

into a primary decomposition. Let $\operatorname{ann} w_{ij} = \left(p_i^{e_{ij}} \right)$. For the convenience of our proof we will assume that and $e_{i1} \geq e_{i2} \geq \cdots \geq e_{it_i} > 0$ for all i and j. (This is the reverse order to what we have been using so far.) Let $\tau = \max\{ t_i : 1 \leq i \leq n \}$. Note that there is at least some i_0 such that $e_{i_0 \tau} \neq 0$. Define $e_{ij} = 0$ if $t_i < j \leq \tau$ so that e_{ij} are defined for all $1 \leq i \leq n$ and $1 \leq j \leq \tau$.

Let $\operatorname{ann} z_k = (d_k)$ for each k. From CRT, we may further decompose $\bigoplus_i D z_i$ into a primary decomposition. This collection of elementary divisor ideals must be the same as the collection in (b). If we reverse the order of the z_i's, which does not truly affect the result of the theorem, we may assume that $d_{k+1} | d_k$ for each k. From Exercise 1, §4.2, we know that in a primary factorization of d_1, the power of p_i must be maximal among the e_{ij}'s. Hence $d_1 \sim p_1^{e_{11}} p_2^{e_{21}} \cdots p_n^{e_{n1}}$. Similarly, $d_k \sim p_1^{e_{1k}} p_2^{e_{2k}} \cdots p_n^{e_{nk}}$ for $1 \leq k \leq \tau$. The argument forces $s = \tau$. We have shown that the invariant factor ideals can be recovered from the elementary divisor ideals. Hence the invariant factor ideals are independent of the decompositions of M. $\quad \square$

From Proposition 4.4.8 and Theorem 4.4.21 we have the following immediate result, which we will leave as an easy exercise. See Exercise 12.

Corollary 4.4.27. *let D be a PID. Two finitely generated modules M and N over D are isomorphic if and only if one of the two following conditions are satisfied.*

- *The modules have the same torsion free rank and invariant factor ideals.*

- *The modules have the same torsion free rank and elementary divisor ideals.*

In Examples 1.1.4 and 1.1.19, we have shown that an abelian group may be seen as a $\mathbb{Z}$-module. We leave it as an easy exercise that a group homomorphism between two abelian groups is also a $\mathbb{Z}$-linear map (see Exercise 5). Hence any problem involving abelian groups may be treated as a problem on $\mathbb{Z}$-modules.

Applying Corollary 4.4.27, we have the following result, known as *Fundamental theorem of finitely generated abelian groups*.

Theorem 4.4.28. *Any finitely generated abelian group is a direct sum of a finite group (its torsion group) and a free abelian group. The rank of the free component is an invariant. Any finite abelian group is a direct sum of cyclic groups of prime power orders. These orders, counted with multiplicities, are uniquely determined and constitute a complete set of invariants in the sense that two finite abelian groups are isomorphic if and only if they have the same set of these invariants.*

Definition 4.4.29. Let R be a ring. The **unit group** of R, denoted $U(R)$, is defined as the set $\{\, u \in R : u \text{ is a unit in } R \,\}$.

Lemma 4.4.30. *Let R be a ring. Then $U(R)$ is abelian as a multiplicative group.*

We leave it as an exercise to verify this lemma, which makes the name "unit group" appropriate. See Exercise 4.

Remember that

$$
\begin{aligned}
U(\mathbb{Z}_n) &= \{\, \overline{k} \in \mathbb{Z}_n \setminus \{0\} : (k, n) \sim 1 \,\} \\
&= \{\, \overline{k} \in \mathbb{Z}_n : 1 \le k \le n,\ (k, n) \sim 1 \,\}
\end{aligned}
$$

for $n \ge 2$.

Definition 4.4.31. The **Euler φ-function**, denoted $\varphi(n)$, is the function from $\mathbb{Z}_+$ to $\mathbb{Z}_+$ defined as follows. Define $\varphi(1) = 1$ and $\varphi(n) = |U(\mathbb{Z}_n)|$ for $n \ge 2$.

For an integer $n \ge 2$, factor $n = p_1^{e_1} p_2^{e_2} \cdots p_n^{e_n}$ where the p_i's are distinct positive prime integers. Then

$$
\varphi(n) = n \left(1 - \frac{1}{p_1} \right) \left(1 - \frac{1}{p_2} \right) \cdots \left(1 - \frac{1}{p_n} \right).
$$

The value of the Euler φ-function gives the number of integers from 1 to n which are relatively prime with n.[2]

---

[2] One may also define $\varphi(1)$ to be 0, in which case the value $\varphi(n)$ gives the number of integers from 1 to $n - 1$ which are relatively prime with n. The two definitions coincide at every integer greater than 1.

Example 4.4.32. In this example we examine the structure of the multiplicative group $U(\mathbb{Z}_{16})$. This group is a finite abelian group of order $\varphi(n) = 16(1/2) = 8$. From Theorem 4.4.28, the group $U(\mathbb{Z}_{16})$ is isomorphic to one of the three following groups

$$\mathbb{Z}_8, \quad \mathbb{Z}_2 \oplus \mathbb{Z}_4, \quad \text{or} \quad \mathbb{Z}_2 \oplus \mathbb{Z}_2 \oplus \mathbb{Z}_2.$$

First, observe the powers of 3 and the powers of 5 in $\mathbb{Z}_{16}$:

$$\text{Powers of 3:} \quad 3, \ 9, \ 11, \ 1, \quad o(3) = 4;$$
$$\text{Powers of 5:} \quad 5, \ 9, \ 13, \ 1, \quad o(5) = 4.$$

There are elements of order 4 in $U(\mathbb{Z}_{16})$, and so $\mathbb{Z}_2 \oplus \mathbb{Z}_2 \oplus \mathbb{Z}_2$ is excluded. In $U(\mathbb{Z}_{16})$ there are at least 4 elements of order 4: 3, 11, 5 and 13. Since there are only 2 elements of order 4 in $\mathbb{Z}_8$, $\mathbb{Z}_8$ is excluded as well. We conclude that $U(\mathbb{Z}_{16}) \cong \mathbb{Z}_4 \oplus \mathbb{Z}_2$. Note that the element $7 \notin \langle 3 \rangle$ and $o(7) = 2$. We leave it to the reader to check that $U(\mathbb{Z}_{16}) = \langle 7 \rangle \langle 3 \rangle$ and $\langle 7 \rangle \cap \langle 3 \rangle = \{1\}$. Thus $U(\mathbb{Z}_{16}) \cong \langle 7 \rangle \times \langle 3 \rangle \cong \mathbb{Z}_2 \oplus \mathbb{Z}_4$.

Exercises 4.4

1. Let D be a PID and let M be the D-module given in (4.4.1). Show that $\operatorname{tor}_D M = Dz_1 \oplus Dz_2 \oplus \cdots \oplus Dz_r$ as claimed in the text.

2. Let M be the $\mathbb{Z}$-module given in Example 4.4.7. Find the number of free submodules L of M such that $M = \operatorname{tor}_{\mathbb{Z}} M \oplus L$.

3. Prove Corollary 4.4.20.

4. Prove Lemma 4.4.30.

5. Let G and G' be two abelian groups. They may also be viewed naturally as $\mathbb{Z}$-modules. Show that a group homomorphisms from G to G' is a $\mathbb{Z}$-linear map. Show that a $\mathbb{Z}$-linear map from G to G' is a group homomorphism.

6. Let R be a ring and let M be an R-module. Suppose $M_1, M_2, \ldots, M_s$ are submodules of M such that

$$M = M_1 \oplus M_2 \oplus \cdots \oplus M_s.$$

Let $m = \sum_{i=1}^{s} m_i$ where $m_i \in M_i$ for all i. Show that

$$\operatorname{ann}_R m = \bigcap_i \operatorname{ann}_R m_i.$$

7. Determine whether the two abelian groups

$$\mathbb{Z}_{72} \oplus \mathbb{Z}_{54} \oplus \mathbb{Z}_5 \quad \text{and} \quad \mathbb{Z}_{18} \oplus \mathbb{Z}_{40} \oplus \mathbb{Z}_{27}$$

are isomorphic.

8. Determine the structure of $U(\mathbb{Z}_{32})$ and of $U(\mathbb{Z}_{100})$.

9. Determine whether the $\mathbb{Z}$-module

$$\frac{\mathbb{Z}^4}{\langle\, (1,\, 3,\, 2,\, 5),\, (2,\, 0,\, 6,\, 6),\, (-3,\, 3,\, -10,\, -6)\,\rangle}$$

is isomorphic to $\dfrac{\mathbb{Z}^3}{\langle\, (2,\, 1,\, 2),\, (4,\, 0,\, 2)\,\rangle}$.

10. Prove Lemma 4.4.26 in the following steps.

    (a) Show that $p^{i+1} M$ is a submodule of $p^i M$ for each i.

    (b) Show that we may view $p^i M / p^{i+1} M$ as a $D/(p)$-module. Remember that pD is a maximal ideal of D. Hence $D/(p)$ is a field, and $p^i M / p^{i+1} M$ is a $D/(p)$-vector space.

    (c) Let $M = D/(p^m)$ and let $F = D/(p)$. Show that

$$\dim_F \frac{p^i M}{p^{i+1} M} = \begin{cases} 1, & \text{if } i < m; \\ 0, & \text{if } i \geq m. \end{cases}$$

    (d) Suppose that $M \cong \dfrac{D}{(p^{j_1})} \oplus \dfrac{D}{(p^{j_2})} \oplus \cdots \oplus \dfrac{D}{(p^{j_s})}$. Show that

$$\dim_F \frac{p^i M}{p^{i+1} M} = n_i = \#\{\, k \in \{1, 2, \ldots, s\} : j_k > i \,\}.$$

    Show that the j_k's can be recovered from the n_i's. Conclude that the j_k's are independent of primary decompositions of M.

11. Let D be a PID and let p, q be distinct primes in D. Suppose M is the cyclic module $D/(q^m)$. Show that $p^i M/p^{i+1} M = 0$ for all $i \geq 0$.

12. Prove Corollary 4.4.27.

4.5 Rational canonical form

Throughout this section V stands for a finite dimensional vector space over the field F. In this section we want to apply the structure theorem on modules to study a linear endomorphism T on V.

Linear endomorphisms and similar matrices

Definition 4.5.1. Let $T\colon V \to V$ be an F-linear transformation. We say that T is a **linear endomorphism** (or a **linear operator**) on V over F.

The set of all linear endomorphisms $\mathrm{Hom}_F(V, V)$ usually forms a *noncommutative* ring (see Exercise 14, §2.1). If $\dim_F V = n$, for any ordered basis β of V, there is a natural ring homomorphism from $\mathrm{Hom}_F(V, V)$ to $M_n(F)$ sending T to the square matrix representing T with respect to β and β. When we study a linear endomorphism, it is reasonable to require the same basis be chosen for both the domain and the codomain.

Assume that $\dim_F V = n$. Let $\beta = (u_1, u_2, \ldots, u_n)$ be an ordered basis of V over F and let $T\colon V \to V$ be the linear endomorphism such that

$$(4.5.1) \qquad Tu_j = \sum_{i=1}^n a_{ij} u_i, \qquad j = 1, \ldots, n.$$

Then $A = (a_{ij})$ is the matrix of T with respect to β and β.

Let $\beta' = (v_1, v_2, \ldots, v_n)$ be another basis such that $v_j = \sum_i p_{ij} u_i$. Then $P = (p_{ij})$ is an invertible matrix by Corollary 2.3.8 since it is the base change matrix from β' to β. The matrix of T with respect to β' and β' is $P^{-1}AP$. Conversely, if P is invertible, P may be viewed as a base change matrix from β' to β by Proposition 4.1.1.

Definition 4.5.2. Let A and B be two square matrices of size n over F. We say that A is **similar** to B if there is an invertible matrix P in $M_n(F)$ such that $B = P^{-1}AP$.

We leave it to the reader to verify that "similarity" on $M_n(F)$ is an equivalence relation. See Exercise 1.

Similar matrices may represent the same linear endomorphism with respect to different bases. The mission now is to find a matrix as simple as possible and the corresponding basis with respect to which the matrix represents a given linear endomorphism.

The F-vector space V as an $F[\lambda]$-module via T

Suppose given a linear endomorphism T on V over F. For the rest of this section λ denotes an indeterminate over F. Consider the polynomial ring $F[\lambda]$ of one variable over the field F. We can make V into an $F[\lambda]$-module by letting

$$(4.5.2) \qquad g(\lambda)v = b_0 v + b_1 T(v) + \cdots + b_m T^m(v)$$

where $g(\lambda) = b_0 + b_1 \lambda + \cdots + b_m \lambda^m \in F[\lambda]$ and $v \in V$. See Exercise 2.

From Exercise 14, §2.1, we have that $F[T]$ is a (commutative) subring of $\mathrm{Hom}_F(V, V)$. The vector space V may also be seen as an $F[T]$-module. Why do we have to introduce a new indeterminate λ? The ring $F[T]$ is never a PID, which makes it difficult to handle. On the other hand, $F[\lambda]$ is a PID. We may apply Theorem 4.3.14 (the structure theorem of finitely generated modules over a PID) to study the structure of V over $F[\lambda]$.

Lemma 4.5.3. *The finite dimensional vector space V over F is a torsion module over $F[\lambda]$.*

Proof. Let $v \in V$. Consider the set $\{T^i v \in V : i \in \mathbb{N}\}$. Since V is finite dimensional over F, we may find $N > 0$ such that $v, Tv, T^2 v, \ldots, T^N v$ are linearly dependent over F by Corollary 1.4.3. Let $a_0, a_1, a_2, \ldots, a_N$ be elements (not all trivial) in F such that $\sum_{i=0}^{N} a_i T^i v = 0$. Hence $g(\lambda)v = 0$ where $g(\lambda) = \sum_{i=0}^{N} a_i \lambda^i$ is a nontrivial polynomial. This shows that any element of V is torsion. $\qquad \square$

Lemma 4.5.4. *Suppose T is the linear endomorphism given in (4.5.1). Let $\{e_i\}_i$ be the standard basis of $F[\lambda]^n$ over $F[\lambda]$. Define*

$$f_j = \lambda e_j - \sum_{i=1}^{n} a_{ij} e_i \in F[\lambda]^n, \qquad 1 \leq j \leq n.$$

Then $\lambda^k e_\ell$ is the sum of an $F[\lambda]$-linear combination of the f_j's and an F-linear combination of the e_i's for all k and ℓ.

Proof. We will prove this lemma by induction on k. Since $\lambda^0 e_\ell = e_\ell$, the lemma holds for $k = 0$. We now assume $k > 0$. From the induction hypothesis, we have $\lambda^{k-1} e_\ell = \sum_{j=1}^n h_j f_j + \sum_{i=1}^n b_i e_i$ where $h_j \in F[\lambda]$ and $b_i \in F$ for all i and j. It follows that

$$
\begin{aligned}
\lambda^k e_\ell &= \lambda \left(\sum_{j=1}^n h_j f_j + \sum_{i=1}^n b_i e_i \right) = \sum_{j=1}^n \lambda h_j f_j + \sum_{i=1}^n b_i \lambda e_i \\
&= \sum_{j=1}^n \lambda h_j f_j + \sum_{i=1}^n b_i \left(f_i + \sum_{r=1}^n a_{ri} e_r \right) \\
&= \sum_{j=1}^n (\lambda h_j + b_j) f_j + \sum_{r=1}^n (b_1 a_{r1} + b_2 a_{r2} + \cdots + b_n a_{rn}) e_r.
\end{aligned}
$$

Thus the lemma holds by induction on k. $\square$

Proposition 4.5.5. *Let $\beta = (u_1, u_2, \ldots, u_n)$ be an ordered basis for V over F. Under the same assumption as in Lemma 4.5.4, we have an R-linear epimorphism $\eta : F[\lambda]^n \to V$ sending e_i to u_i for $1 \le i \le n$. The elements $f_j = \lambda e_j - \sum_{i=1}^n a_{ij} e_i$, $1 \le j \le n$, generate $\mathrm{Ker}\,\eta$ over $F[\lambda]$.*

Proof. The part that η is surjective follows from the fact that the basis β spans V over $F[\lambda]$. We proceed to find $\mathrm{Ker}\,\eta$.

For each j, it is clear that $f_j = \lambda e_j - \sum_{i=1}^n a_{ij} e_i \in \mathrm{Ker}\,\eta$ since

$$
\eta(f_j) = \lambda \eta(e_j) - \sum_1^n a_{ij} \eta(e_i) = T u_j - \sum_1^n a_{ij} u_i = 0
$$

by (4.5.1). Now let $\sum_{i=1}^n g_i(\lambda) e_i \in \mathrm{Ker}\,\eta$, we need to show that it is generated by the f_j's over $F[\lambda]$. Note that $\sum_{i=1}^n g_i(\lambda) e_i$ is an F-linear combination of the $\lambda^k e_\ell$'s. From Lemma 4.5.4 we have

$$
\sum_{i=1}^n g_i(\lambda) e_i = \sum_{j=1}^n h_j f_j + \sum_{i=1}^n b_i e_i
$$

where $h_j \in F[\lambda]$ and $b_i \in F$ for all i and j. It follows that

$$
\eta \left(\sum_{i=1}^n g_i(\lambda) e_i \right) = \sum_{i=1}^n b_i u_i = 0.
$$

This implies that $b_i = 0$ for all i since the u_i's are linearly independent over F. Hence $\sum_{i=1}^{n} g_i(\lambda)e_i$ is generated by the f_j's over $F[\lambda]$. $\square$

From Proposition 4.5.5, Ker η is the column space of

$$\lambda I - A = \begin{pmatrix} \lambda - a_{11} & -a_{12} & \cdots & -a_{1n} \\ -a_{21} & \lambda - a_{22} & \cdots & -a_{2n} \\ \multicolumn{4}{c}{\dotfill} \\ \multicolumn{4}{c}{\dotfill} \\ -a_{n1} & -a_{n2} & \cdots & \lambda - a_{nn} \end{pmatrix}.$$

Thus as an $F[\lambda]$-module, we have $V \cong \text{Coker}\,(\lambda I - A)$. The normal form of $\lambda I - A$ will give us the structure of V as a module over $F[\lambda]$ via T.

Definition 4.5.6. Let A be a square matrix over a ring. We call

$$f(\lambda) = \det(\lambda I - A) = \lambda^n - b_1\lambda^{n-1} + \cdots + (-1)^n b_n$$

the **characteristic polynomial** of A. The coefficient $b_1 = \sum_i a_{ii}$ is called the **trace** of A.

Note that $b_n = \det A$.

Let the normal form of $\lambda I - A \in M_n(F[\lambda])$ be

$$(4.5.3) \qquad\qquad \text{diag}\{1, \ldots, 1, d_1(\lambda), \ldots, d_s(\lambda)\},$$

where the d_i's are of positive degree and $d_i(\lambda)|d_j(\lambda)$ for $i < j$. This is possible since we have that $d_s(\lambda) \neq 0$ by Lemma 4.5.3. We may further assume that all the d_i's are monic. The determinant of the normal form is $d_1(\lambda) \cdots d_s(\lambda)$, which is an associate to the determinate of $\lambda I - A$. By comparing the leading coefficients we can see that the characteristic polynomial $f(\lambda)$ of A is indeed $d_1(\lambda) \cdots d_s(\lambda)$. From the structure theorem we have that

$$(4.5.4) \quad V = F[\lambda]z_1 \oplus F[\lambda]z_2 \oplus \cdots \oplus F[\lambda]z_s, \quad \text{where ann } z_i = (d_i(\lambda)).$$

Since the module structure of V depends only on T, not on the basis chosen, we can also define the characteristic polynomial of T to be $d_1(\lambda) \cdots d_s(\lambda)$. *Cf.* Exercise 3.

Definition 4.5.7. The polynomial $m(\lambda) = d_s(\lambda)$ in (4.5.3) is called the **minimal polynomial** of A (or of T).

This terminology comes from the following result due to Hamilton, Cayley and Frobenius.

Theorem 4.5.8 (Cayley-Hamilton theorem). *Let T be a linear endomorphism on a finite dimensional F-vector space. Let $f(\lambda)$ and $m(\lambda)$ be the characteristic polynomial and minimal polynomial of T respectively. The following statements are true.*

(a) *The minimal polynomial $m(\lambda)$ is divisor of $f(\lambda)$ in $F[\lambda]$.*

(b) *The two polynomials $m(\lambda)$ and $f(\lambda)$ have the same prime factors in $F[\lambda]$.*

(c) *Let $g(\lambda) \in F[\lambda]$. Then $g(T) = 0$ if and only if $m(\lambda)|g(\lambda)$ in $F[\lambda]$.*

The statements above are also true for any matrix representing T.

Part (c) explains why we call $m(\lambda)$ the minimal polynomial of T.

Proof. (a) Using the notation in (4.5.3) we have

$$m(x) = d_s(\lambda) \,|\, d_1(\lambda) \cdots d_s(\lambda) = f(\lambda).$$

(b) A prime factor of $m(\lambda)$ is a prime factor of $f(\lambda)$ by (a). If $p(\lambda)$ is a prime factor of $f(\lambda)$, it is a prime divisor of $d_i(\lambda)$ for some i. Since $d_i(\lambda)$ is a factor of $d_s(\lambda)$, $p(\lambda)$ is a divisor of $m(\lambda)$.

(c) We will use the notation in (4.5.4).

The "if" part: Let $g(\lambda) = m(\lambda)h(\lambda)$. Note that $m(\lambda)z_i = 0$ for all i since $m(\lambda) \in \operatorname{ann} z_i$ for all i. Let $v \in V$. Find $h_1, h_2, \ldots, h_n \in F[\lambda]$ such that $v = \sum_i h_i(\lambda)z_i$. Then

$$m(\lambda)v = m(\lambda)\sum_i h_i(\lambda)z_i = \sum_i h_i(\lambda)m(\lambda)z_i = 0.$$

It follows that $g(T)v = h(\lambda)m(\lambda)v = h(\lambda)0 = 0$ for all $v \in V$. In other words, $g(T) = 0$.

The "only if" part: Let $g(T) = 0$. Then $g(T)z_s = g(\lambda)z_s = 0$. This implies that $g(\lambda) \in \operatorname{ann} z_s = \big(m(\lambda)\big)$. Thus $m(\lambda)\,|\,g(\lambda)$ in $F[\lambda]$. $\square$

Subspaces stabilized by T

Definition 4.5.9. Let $T\colon V \to V$ be an F-linear endomorphism and let W be an F-subspace of V. We say that W is **stabilized** by T if $T(W) \subseteq W$. Or we can say that W is a T-**invariant** subspace of V.

The following lemma gives a new description to T-invariant subspaces.

Lemma 4.5.10. *Let T be an F-linear endomorphism on V and this makes V into an $F[\lambda]$-module via T. The following statements are true.*

(a) *Any $F[\lambda]$-submodule of V is an F-subspace of V.*

(b) *An F-subspace W of V is stabilized by T if and only if W is an $F[\lambda]$-submodule of V.*

Proof. We will prove the "if" part of (b) and leave the rest as an easy exercise to the reader. See Exercise 7.

Let W be an $F[\lambda]$-submodule of V. Then $T(w) = \lambda w \in W$. Hence W is stabilized by T. $\square$

When W is a subspace of V stabilized by the linear endomorphism $T\colon V \to V$. We may restrict the domain and codomain of T to obtain a new linear endomorphism as follows

$$(4.5.5) \qquad \begin{array}{rccc} T_W\colon & W & \longrightarrow & W \\ & w & \longmapsto & T(w) \end{array}$$

Suppose we may find two subspaces U and W of V which are stabilized by an endomorphism T such that $V = U \oplus W$. Find an ordered basis $(u_1, u_2, \ldots, u_\ell)$ for U and an ordered basis $(w_1, w_2, \ldots, w_m)$ for W. Then

$$\beta = (u_1, u_2, \ldots, u_\ell, w_1, w_2, \ldots, w_m)$$

is an ordered basis for V over F (see Exercise 9, §1.4). Let A be the matrix representing $T_U\colon U \to U$ with respect to $(u_1, u_2, \ldots, u_\ell)$ and B be the matrix representing $T_W\colon W \to W$ with respect to $(w_1, w_2, \ldots, w_m)$. It is easy to see that the matrix representing $T\colon V \to V$ with respect to β is

$$\begin{pmatrix} A & \\ & B \end{pmatrix}.$$

To analyze T it is sufficient to analyze A and B, which greatly simplifies matter. Hence our goal is to break V into as many T-invariant subspaces as possible. To analyze a matrix representing T is thus equivalent to analyzing several smaller matrices.

Remember that the linear endomorphism T makes V an $F[\lambda]$-module. In (4.5.4) the vector space V is a direct sum of the cyclic modules $F[\lambda]z_i$. From the "if" part of Lemma 4.5.10(b), we have that $F[\lambda]z_i$ is an F-subspace stabilized by T inside V. To understand T on V is thus to understand $T_{F[\lambda]z_i}$ for all i. For the following discussion we will thus concentrate on the case of nontrivial cyclic torsion modules over $F[\lambda]$.

Rational form

We are ready now to describe the first type of canonical form for a square matrix over a PID.

Lemma 4.5.11. *Let* $\deg d(\lambda) = n$. *The set* $S = \{\overline{1}, \overline{\lambda}, \overline{\lambda}^2, \ldots, \overline{\lambda}^{n-1}\}$ *is an F-basis for $F[\lambda]/(d(\lambda))$. Moreover, the set* $\{z, \lambda z, \lambda^2 z, \ldots, \lambda^{n-1}z\}$ *is an F-basis for $F[\lambda]z$ if* $\operatorname{ann} z = (d(\lambda))$.

Proof. Let $g(\lambda) \in F[\lambda]$. Find $q(\lambda), r(\lambda) \in F[\lambda]$ such that

$$g(\lambda) = q(\lambda)d(\lambda) + r(\lambda), \qquad \text{where } r(\lambda) = 0 \text{ or } \deg r(\lambda) < \deg d(\lambda).$$

Then $\overline{g(\lambda)} = \overline{r(\lambda)}$ is generated by S over F. On the other hand, let

$$a_0 \cdot \overline{1} + a_1 \cdot \overline{\lambda} + \cdots + a_1 \cdot a_{n-1}\overline{\lambda}^{n-1} = \overline{a_0 + a_1\lambda + \cdots + a_{n-1}\lambda^{n-1}} = 0.$$

Then $f(\lambda) = a_0 + a_1\lambda + \cdots + a_{n-1}\lambda^{n-1} = d(\lambda)q(\lambda)$ for some $q(\lambda) \in F[\lambda]$. By comparing the degrees on both sides, we see that $f(\lambda)$ must be the trivial polynomial. This implies that $a_i = 0$ for all i. It follows that S is linearly independent over F. Hence S is an F-basis for $F[\lambda]/(d(\lambda))$.

From the proof of Lemma 4.3.2 we deduce that we may identify $\lambda^i z$ in $F[\lambda]z$ with $\overline{\lambda}^i$ in $F[\lambda]z/(d(\lambda))$. This shows that the second part of the lemma is also true. $\qquad\square$

Proposition 4.5.12. *Let T be an F-linear endomorphism on V such that* $V = F[\lambda]z$ *where* $\operatorname{ann} z$ *is the ideal generated by*

$$d(\lambda) = \lambda^n + b_{n-1}\lambda^{n-1} + b_{n-2}\lambda^{n-2} + \cdots + b_1\lambda + b_0.$$

The matrix representing T with respect to $\beta = (z, \lambda z, \lambda^2 z, \ldots, \lambda^{n-1} z)$ is

$$(4.5.6) \qquad B = \begin{pmatrix} 0 & 0 & 0 & \cdots & 0 & 0 & -b_0 \\ 1 & 0 & 0 & \cdots & 0 & 0 & -b_1 \\ 0 & 1 & 0 & \cdots & 0 & 0 & -b_2 \\ & & & \cdots\cdots\cdots\cdots\cdots & & \\ & & & \cdots\cdots\cdots\cdots\cdots & & \\ 0 & 0 & 0 & \cdots & 1 & 0 & -b_{n-2} \\ 0 & 0 & 0 & \cdots & 0 & 1 & -b_{n-1} \end{pmatrix}.$$

*The matrix B is called the **companion matrix** of the polynomial $d(\lambda)$.*

Proof. Remember that the isomorphism $F[\lambda]/(d(\lambda)) \longrightarrow F[\lambda]z$ is given by corresponding $\overline{g(\lambda)}$ with $g(\lambda)z$. From Lemma 4.5.11 we have that β is an ordered basis for $F[\lambda]z$. Applying T we have

$$\begin{aligned} z &\longmapsto \lambda z \\ \lambda z &\longmapsto \lambda^2 z \\ &\vdots \\ \lambda^{n-2} z &\longmapsto \lambda^{n-1} z \\ \lambda^{n-1} z &\longmapsto \lambda^n z = -b_0 z - b_1 \lambda z - \cdots - b_{n-1} \lambda^{n-1} z \end{aligned}$$

since $d(\lambda)z = 0$. Thus B is the matrix representing T with respect to β. $\square$

Using Proposition 4.5.12, we now have the following main result.

Theorem 4.5.13. *Let T be an F-linear endomorphism on V. Let $d_i(\lambda)$, $i = 1, 2, \ldots, s$, be the invariant factors of V via T satisfying the requirement $d_i \mid d_{i+1}$ for each i. Then there exist $z_1, z_2, \ldots, z_s$ in V such that*

$$V = F[\lambda]z_1 \oplus F[\lambda]z_2 \oplus \cdots \oplus F[\lambda]z_s$$

where $\operatorname{ann} z_i = (d_i(\lambda))$ *and* $\deg d_i(\lambda) = n_i$ *for each i. The matrix representing T with respect to the ordered basis*

$$(z_1, \lambda z_1, \ldots, \lambda^{n_1 - 1} z_1; z_2, \lambda z_2, \ldots, \lambda^{n_2 - 1} z_2; \cdots; z_s, \lambda z_s, \cdots, \lambda^{n_s - 1} z_s)$$

is of the form

$$(4.5.7) \qquad B = \begin{pmatrix} B_1 & & & \\ & B_2 & & \\ & & \ddots & \\ & & & B_s \end{pmatrix}$$

where B_i is the companion matrix of $d_i(\lambda)$.

The matrix B in (4.5.7) is called the **rational canonical form**, or simply the **rational form**, of the linear endomorphism T (or of any matrix representing T).

Example 4.5.14. Let T be the linear endomorphism on $V = \mathbb{Q}^3$ such that

$$
\begin{aligned}
Tu_1 &= 4u_1 - u_2 - 2u_3 \\
Tu_2 &= -3u_1 + 2u_2 + 2u_3 \\
Tu_3 &= 6u_1 - 2u_2 - 3u_3
\end{aligned}
$$

Find the rational canonical form B of T. Let

$$
A = \begin{pmatrix} 4 & -3 & 6 \\ -1 & 2 & -2 \\ -2 & 2 & -3 \end{pmatrix}.
$$

Find an invertible matrix P over $\mathbb{Q}$ such that $B = P^{-1}AP$.

Solution. First we need to diagonalize the matrix

$$
\lambda I - A = \begin{pmatrix} \lambda - 4 & 3 & -6 \\ 1 & \lambda - 2 & 2 \\ 2 & -2 & \lambda + 3 \end{pmatrix}.
$$

While doing so we also need to record all the elementary row operations used in order to find P later.

Step 1. Switch the first and the second row of $\lambda I - A$:

$$
\lambda I - A \rightsquigarrow A_1 = E_1(\lambda I - A) = \begin{pmatrix} 1 & \lambda - 2 & 2 \\ \lambda - 4 & 3 & -6 \\ 2 & -2 & \lambda + 3 \end{pmatrix}
$$

where

$$
E_1 = \begin{pmatrix} 0 & 1 & \\ 1 & 0 & \\ & & 1 \end{pmatrix}.
$$

Step 2. Add $(-\lambda + 4) \times$row 1 to row 2, and then add $(-2) \times$row 1 to row 3:

$$
A_1 \rightsquigarrow A_2 = E_2 A_1 = \begin{pmatrix} 1 & \lambda - 2 & 2 \\ 0 & -\lambda^2 + 6\lambda - 5 & -2\lambda + 2 \\ 0 & -2\lambda + 2 & \lambda - 1 \end{pmatrix}
$$

where

$$E_2 = \begin{pmatrix} 1 & 0 & 0 \\ -(\lambda - 4) & 1 & 0 \\ -2 & 0 & 1 \end{pmatrix}.$$

Step 3. Use the pivot in the first row to cancel everything else in the first row. This requires two elementary column operations:

$$A_2 \rightsquigarrow A_3 = \begin{pmatrix} 1 & 0 & 0 \\ 0 & -\lambda^2 + 6\lambda - 5 & -2\lambda + 2 \\ 0 & -2\lambda + 2 & \lambda - 1 \end{pmatrix}.$$

Step 4. Switch column 2 and column 3 of A_3:

$$A_3 \rightsquigarrow A_4 = \begin{pmatrix} 1 & 0 & 0 \\ 0 & -2\lambda + 2 & -\lambda^2 + 6\lambda - 5 \\ 0 & \lambda - 1 & -2\lambda + 2 \end{pmatrix}.$$

Step 5. Switch row 2 and row 3 of A_4:

$$A_4 \rightsquigarrow A_5 = E_3 A_4 = \begin{pmatrix} 1 & 0 & 0 \\ 0 & \lambda - 1 & -2\lambda + 2 \\ 0 & -2\lambda + 2 & -\lambda^2 + 6\lambda - 5 \end{pmatrix},$$

where

$$E_3 = \begin{pmatrix} 1 & & \\ & 0 & 1 \\ & 1 & 0 \end{pmatrix}.$$

Step 6. Add 2×column 2 to column 3:

$$A_5 \rightsquigarrow A_6 = \begin{pmatrix} 1 & & \\ & \lambda - 1 & 0 \\ & -2(\lambda - 1) & -(\lambda - 1)^2 \end{pmatrix}.$$

Step 7. Add 2×row 2 to row 3:

$$A_6 \rightsquigarrow A_7 = E_4 A_6 = \begin{pmatrix} 1 & & \\ & \lambda - 1 & 0 \\ & 0 & -(\lambda - 1)^2 \end{pmatrix}$$

where

$$E_4 = \begin{pmatrix} 1 & & \\ & 1 & 0 \\ & 2 & 1 \end{pmatrix}.$$

Step 8. Multiply the third column of A_7 by (-1):

(4.5.8) $$A_7 \rightsquigarrow A_8 = \begin{pmatrix} 1 & & \\ & \lambda - 1 & 0 \\ & 0 & (\lambda - 1)^2 \end{pmatrix}.$$

In eight steps we have reached the normal form A_8 of $\lambda I - A$. Note as well that $A_8 = Q(\lambda I - A)Q'$ where $Q = E_4 E_3 E_2 E_1$ and Q' is the product of a few elementary matrices coming from the elementary column operations which we did not bother to record. Remember that V is isomorphic to the cokernel of $\lambda I - A$. We may view Q as the base change matrix from the standard basis of $\mathbb{Q}[\lambda]^3$ to another basis β. Note that

$$Q^{-1} = E_1^{-1} E_2^{-1} E_3^{-1} E_4^{-1}$$

$$= \begin{pmatrix} 0 & 1 & \\ 1 & 0 & \\ & & 1 \end{pmatrix} \begin{pmatrix} 1 & 0 & 0 \\ \lambda - 4 & 1 & 0 \\ 2 & 0 & 1 \end{pmatrix} \begin{pmatrix} 1 & & \\ & 0 & 1 \\ & 1 & 0 \end{pmatrix} \begin{pmatrix} 1 & & \\ & 1 & 0 \\ & -2 & 1 \end{pmatrix}$$

$$= \begin{pmatrix} \lambda - 4 & -2 & 1 \\ 1 & 0 & 0 \\ 2 & 1 & 0 \end{pmatrix}.$$

Thus β is the ordered basis

$$\bigl(f_1 = (\lambda - 4, 1, 2), \ f_2 = (-2, 0, 1), \ f_3 = (1, 0, 0)\bigr).$$

We have that $V = \mathbb{Q}[\lambda]z_1 \oplus \mathbb{Q}[\lambda]z_2$ where $z_1 = -2u_1 + u_3$ and $z_2 = u_1$.

From (4.5.8) we have that the invariant factors of V via T are $\lambda - 1$ and $(\lambda - 1)^2 = \lambda^2 - 2\lambda + 1$. This tells us that the rational form of A is

$$B = \left(\begin{array}{c|cc} 1 & & \\ \hline & 0 & -1 \\ & 1 & 2 \end{array} \right).$$

We know that the matrix representing T with respect to

$$\gamma = (z_1, z_2, \lambda z_2) = (-2u_1 + u_3, u_1, 4u_1 - u_2 - 2u_3)$$

is B by Theorem 4.5.13. Thus $B = P^{-1}AP$ where

$$P = \begin{pmatrix} -2 & 1 & 4 \\ 0 & 0 & -1 \\ 1 & 0 & -2 \end{pmatrix}$$

is the base change matrix from γ to (u_1, u_2, u_3).   ◇

Example 4.5.15. Suppose A is a square matrix over $\mathbb{Q}$ such that

$$A^5 - 4A - 2 = 0.$$

Show that A is an invertible matrix of size a multiple of 5.

Solution. Let $g(\lambda) = \lambda^5 - 4\lambda - 2 \in \mathbb{Q}[\lambda]$. The minimal polynomial of A is a divisor of $g(\lambda)$. Since the polynomial $g(\lambda)$ is irreducible over $\mathbb{Q}$ by Eisenstein's criterion,[3] the minimal polynomial of A must be $g(\lambda)$. The invariant factors of $\mathrm{Coker}\,(\lambda I - A)$ are therefore a series of $g(\lambda)$. This implies that the rational form of A consists of a number of the companion matrix of $g(\lambda)$ which is of size 5. Hence the size of A, which is the size of its rational form, is a multiple of 5. Note that the determinant of the companion matrix of $g(\lambda)$ is equal to 2 (see Exercise 6). The determinant of A, which is equal to the determinant of its rational form, is a power of 2 (see Exercise 3). We conclude that A is an invertible matrix of size a multiple of 5.   ◇

Exercises 4.5

Throughout these exercises, F denotes a field and λ denotes an indeterminate over F or over $\mathbb{Q}$.

1. Show that the "similarity" defined in Definition 4.5.2 is an equivalence relation.

---

[3] Here is a simple form of Eisenstein's criterion. Let $f(x) \in \mathbb{Z}[x]$. If there is a prime integer p such that p does not divide the leading coefficient, p divides all other coefficients and p^2 does not divide the constant term of $f(x)$, then $f(x)$ is irreducible over $\mathbb{Q}$. This is a result discussed in an undergraduate course on abstract algebra.

2. Show that the action in (4.5.2) makes V into an $F[\lambda]$-module via T.

3. Show that $\det A = \det B$ and $\det(\lambda I - A) = \det(\lambda I - B)$ for similar matrices A and B in $M_n(F)$ by direct computation.

4. Let A and B be square matrices of the same size over F. Show that the matrices A and B are similar over F if and only if the matrices $\lambda I - A$ and $\lambda I - B$ are equivalent over $F[\lambda]$.

5. Show that any square matrix A over F is similar to its transpose A^{tr}.

6. Show that the determinant of the companion matrix in (4.5.6) is equal to $(-1)^n b_0$ by direct computation.

7. Finish the proof for Lemma 4.5.10.

8. The matrix
$$\left(\begin{array}{cc|ccc} 0 & 1 & & & \\ 1 & 0 & & & \\ \hline & & 0 & 0 & 1 \\ & & 1 & 0 & 0 \\ & & 0 & 1 & 0 \end{array}\right) \in M_5(F)$$

is composed of two blocks of companion matrices. However, it is not a rational form. Find its rational form.

9. Find a rational form for the matrix
$$\left(\begin{array}{cccc} 1 & 0 & 0 & 0 \\ 0 & 1 & 0 & 0 \\ -2 & -2 & 0 & 1 \\ -2 & 0 & -1 & -2 \end{array}\right) \in M_4(\mathbb{Q}).$$

10. Let
$$A = \left(\begin{array}{cccccc} 4 & -3 & 9 & 0 & -3 & 0 \\ 0 & 1 & 0 & 0 & 1 & 0 \\ -1 & 1 & -2 & 0 & 1 & 0 \\ 1 & 0 & 3 & 1 & -1 & 0 \\ 0 & 0 & 0 & 0 & 1 & 0 \\ 0 & 0 & 1 & 1 & 0 & 1 \end{array}\right) \in M_6(\mathbb{Q}).$$

(a) Find the normal form of $\lambda I - A$.

(b) Determine the characteristic polynomial and the minimal polynomial of A.

(c) Determine the rational form B of the matrix A.

(d) Find an invertible matrix P such that $B = P^{-1}AP$.

11. Let

$$A = \begin{pmatrix} -1 & 0 & -3 & -3 & 0 & 3 \\ 0 & -1 & 0 & 0 & 1 & 0 \\ 0 & 0 & 0 & 1 & 0 & -1 \\ 1 & 0 & 3 & -1 & -1 & 0 \\ 0 & 0 & 0 & 0 & -1 & 0 \\ -1 & 0 & -2 & 1 & 1 & 0 \end{pmatrix} \in M_6(\mathbb{Q}).$$

(a) Find the normal form of $\lambda I - A$.

(b) Determine the characteristic polynomial and the minimal polynomial of A.

(c) Determine the rational form B of the matrix A.

(d) Find an invertible matrix P such that $B = P^{-1}AP$.

12. Let $d(\lambda) \in F[\lambda]$ be a monic polynomial of positive degree n. If the companion matrix of $d(\lambda)$ represents an linear endomorphism T on the n-dimensional vector space V over F, show that via T, V is a cyclic $F[\lambda]$-module whose order ideal is $(d(\lambda))$.

13. Show that an F-vector space V viewed as an $F[\lambda]$-module via a linear endomorphism T is cyclic if and only if the characteristic polynomial of T is the same as its minimal polynomial.

14. We say that an F-linear endomorphism T on V is a **semisimple operator** if for every T-invariant subspace W of V, there exists a T-invariant subspace U such that $V = W \oplus U$.

Let T be a semisimple operator and let W be a T-invariant subspace of V. Show that T_W (as in (4.5.5)) is semisimple.

15. Let $d(\lambda) \in F[\lambda]$ be monic and of positive degree. Let B be the companion matrix of $d(\lambda)$.

(a) Show that $d(\lambda)$ is the characteristic polynomial of B by computing $\det(\lambda I - B)$ directly.

(b) Let $g(\lambda) \in F[\lambda]$ be a nonzero polynomial of degree less than the degree of $d(\lambda)$. Show that $g(B)e_1 \neq 0$ where e_1 denotes the column n-vector with 1 as the first entry and 0 elsewhere. Use this fact to show that $d(\lambda)$ is also the minimal polynomial of B.

16. Let $f(\lambda) = \lambda^n - b_1\lambda^{n-1} + b_2\lambda^{n-2} - b_3\lambda^{n-3} + \cdots + (-1)^n b_n$ be the characteristic polynomial of A. Show that b_i is the sum of the principal i-minors of A. (Principal i-minors are the i-minors obtained from the submatrix corresponding to the rows $r_1, r_2, \ldots, r_i$ and to the columns $r_1, r_2, \ldots, r_i$ where $1 \leq r_1 < r_2 < \cdots < r_i \leq n$.)

4.6   Jordan canonical form

There is possibly a second type of canonical form for a linear endomorphism. This new type of canonical form is usually "simpler" than the rational canonical form.

Throughout this section, F denotes a field and V denotes a finite dimensional vector space over F.

Jordan form

Definition 4.6.1. Let x be an indeterminate over the field F. We say a polynomial **splits** over F if it is a product of linear polynomials in $F[x]$.

A field F is **algebraically closed** if in $F[x]$ every polynomial of positive degree splits over F.

Fundamental theorem of algebra tells us that the field of complex numbers $\mathbb{C}$ is algebraically closed, while the field of real numbers $\mathbb{R}$ is not.

We summarize what we know about splitting polynomials and splitting characteristic polynomials. We leave Lemma 4.6.2 as a review exercise for the reader. See Exercise 1.

Lemma 4.6.2. *A polynomial splits over the field F if and only if all of its prime factors over F are linear.*

Lemma 4.6.3. *Let T be an F-linear endomorphism on V. The following statements are equivalent.*

(i) *The characteristic polynomial of T splits over F.*

(ii) *All the invariant factors of V via T split over F.*

(iii) *The minimal polynomial of A splits over F.*

(iv) *All the elementary divisors of V via T are of the form $(\lambda - r)^e$ for some $r \in F$ and $e \in \mathbb{Z}_+$.*

Proof. "(i) $\Rightarrow$ (ii)": The characteristic polynomial of T is a product of the invariant factors of V via T. If all the prime factors of the characteristic polynomial of T are linear, so are the prime factors of all the invariant factors of V via T.

"(ii) $\Rightarrow$ (iii)": This follows from the fact that the minimal polynomial of T is one of the invariant factors of V via T.

"(iii) $\Rightarrow$ (iv)": Each elementary divisor of V via T is a power of a prime factor of the characteristic polynomial of A. From Theorem 4.5.8, the prime factors of the characteristic polynomial of T are exactly the prime factors of the minimal polynomial of T. Hence all the prime factors of the characteristic polynomial are linear.

"(iv) $\Rightarrow$ (i)": The characteristic polynomial of T is a product of all the elementary divisors of V via T (repeating elementary divisors must be counted as well), and each elementary divisors is a power of $\lambda - r$ for some $r \in F$. $\qquad\square$

Let's review the following definition.

Definition 4.6.4. Let $T \colon V \to V$ be an F-linear endomorphism. If v is a nonzero vector in V such that $Tv = rv$, we say that r is an **eigenvalue** of T and v is an **eigenvector** of T. Let r be an eigenvalue of T. The subset

(4.6.1) $$E_r = \{v \in V : Tv = rv\} = \mathrm{Ker}\,(T - \mathbf{1}_V)$$

is a subspace of V. The subspace E_r is called the **eigenspace** of T associated with the eigenvalue r.

Let's first look at the cyclic case when $V = F[\lambda]w$ via the F-linear endomorphism T and let ann $w = ((\lambda - r)^e)$ for some $r \in F$ and some positive integer e. The characteristic polynomial $f(\lambda)$ of T equals the minimal polynomial of T and also equals its sole elementary divisor $(\lambda - r)^e$. Using Exercise 2 we have that

$$\beta = (\, w, \ (\lambda - r)w, \ (\lambda - r)^2 w, \ \ldots, \ (\lambda - r)^{e-1}w \,)$$

is an ordered basis for V over F. Note that

$$
\begin{aligned}
w &\longmapsto \lambda w = rw + (\lambda - r)w \\
(\lambda - r)w &\longmapsto \lambda(\lambda - r)w = r(\lambda - r)w + (\lambda - r)^2 w \\
&\ \ \vdots \\
(\lambda - r)^{e-2}w &\longmapsto \lambda(\lambda - r)^{e-2}w = r(\lambda - r)^{e-2}w + (\lambda - r)^{e-1}w \\
(\lambda - r)^{e-1}w &\longmapsto \lambda(\lambda - r)^{e-1}w = r(\lambda - r)^{e-1}w + (\lambda - r)^e w \\
&\qquad\qquad\qquad\quad = r(\lambda - r)^{e-1}w.
\end{aligned}
$$

Thus the matrix of T with respect to β is

(4.6.2)
$$
J = \begin{pmatrix}
r & 0 & 0 & \ldots\ldots & & 0 \\
1 & r & 0 & 0 & \ldots & 0 \\
0 & 1 & r & 0 & \ldots & 0 \\
\multicolumn{6}{c}{\ldots\ldots\ldots\ldots\ldots} \\
\multicolumn{6}{c}{\ldots\ldots\ldots\ldots\ldots} \\
0 & & \ldots\ldots & & 1 & r & 0 \\
0 & & \ldots\ldots\ldots & & & 1 & r
\end{pmatrix}_{e \times e}.
$$

Note that r is an eigenvalue of T and that $(\lambda - r)^{e-1}w$ is an eigenvector of T.

Definition 4.6.5. The matrix in (4.6.2) is called the **Jordan block** with respect to (the eigenvalue) r of size e.

From the discussion above we may conclude with the following theorem.

Theorem 4.6.6. *Let V be a finite dimensional F-vector space and let $T : V \to V$ be an F-linear endomorphism. Suppose the elementary divisors (with multiplicity) of V via T are $(\lambda - r_i)^{e_i}$, $i = 1,\ldots,t$. There exists $w_1, w_2, \ldots, w_t \in V$ such that*

$$V = F[\lambda]w_1 \oplus F[\lambda]w_2 \oplus \cdots \oplus F[\lambda]w_t$$

where ann $w_i = \left((\lambda - r_i)^{e_i}\right)$. *The matrix of T with respect to the ordered basis*

$$\left(w_1, (\lambda - r_1)w_1, \ldots, (\lambda - r_1)^{e_1-1}w_1; w_2, (\lambda - r_2)w_2, \ldots, \right.$$
$$\left. (\lambda - r_2)^{e_2-1}w_2; \ldots; w_t, (\lambda - r_t)w_t, \ldots, (\lambda - r_t)^{e_t-1}w_t \right)$$

is the matrix

(4.6.3)
$$J = \begin{pmatrix} J_1 & & & \\ & J_2 & & \\ & & \ddots & \\ & & & J_t \end{pmatrix}$$

where J_i is the Jordan block with respect to r_i of size e_i.

The matrix J is called a **Jordan canonical form**, or simply a **Jordan form**, of T (or of any matrix representing T).

Corollary 4.6.7. *Let V be a finite dimensional F-vector space and let $T: V \to V$ be an F-linear endomorphism. There is a Jordan form for T if and only if the characteristic polynomial of T splits over F.*

Proof. The "if" part: If the characteristic polynomial of T splits over F, the elementary divisors of V via T are all powers of monic linear polynomials in $F[\lambda]$ by Lemma 4.6.3. The result follows from Theorem 4.6.6.

The "only if" part: Suppose a matrix of the form in (4.6.3) is a Jordan form for T. Note that the characteristic polynomial of T is the product of the characteristic polynomials of the J_i's. We leave the rest of the proof as an exercise. See Exercise 4. $\qquad\square$

Since $\mathbb{C}$ is algebraically closed, a matrix over $\mathbb{C}$ always has a Jordan form. However, a matrix over $\mathbb{R}$ may not possess a Jordan form.

Example 4.6.8. Find a rational form and a Jordan form (if any) for the matrix

$$A = \begin{pmatrix} 0 & 2 \\ -2 & 0 \end{pmatrix}$$

over $\mathbb{R}$ and over $\mathbb{C}$.

Solution. The characteristic polynomial of A is $f(\lambda) = \lambda^2 + 4$.

The case over $\mathbb{C}$: It is obvious that $\lambda^2 + 4$ is the sole invariant factor of Coker $(\lambda I - A)$, and $\lambda + 2i$ and $\lambda - 2i$ are the only two elementary divisors of Coker $(\lambda I - A)$. Hence

$$B = \begin{pmatrix} 0 & -4 \\ 1 & 0 \end{pmatrix} \quad \text{and} \quad J = \begin{pmatrix} -2i & 0 \\ 0 & 2i \end{pmatrix}$$

are a rational form and a Jordan form for A respectively.

The case over $\mathbb{R}$: Since the polynomial $f(\lambda)$ is prime in $\mathbb{R}[\lambda]$, the polynomial $\lambda^2 + 4$ is the sole invariant factor and the sole elementary divisor of Coker $(\lambda I - A)$. Hence B is a rational form for A over $\mathbb{R}$ while A possesses no Jordan form over $\mathbb{R}$. $\diamond$

Example 4.6.9. Let T and A be as in Example 4.5.14. Find a Jordan form J for T and find an invertible matrix P over $\mathbb{Q}$ such that $J = P^{-1}AP$.

Solution. We will follow the notation in Example 4.5.14. From (4.5.8), we derived that the invariant factors of V via T are $\lambda - 1$ and $(\lambda - 1)^2$. They happen to be the elementary divisors of V via T as well. Hence

$$J = \left(\begin{array}{c|cc} 1 & & \\ \hline & 1 & \\ & 1 & 1 \end{array} \right)$$

is a Jordan form for A over $\mathbb{Q}$.

Next we need to find an invertible matrix P such that $J = P^{-1}AP$. In Example 4.5.14 we have found that

$$V = \mathbb{Q}[\lambda]z_1 \oplus \mathbb{Q}[\lambda]z_2, \qquad \text{where } z_1 = -2u_1 + u_3 \text{ and } z_2 = u_1.$$

Note that $\lambda z_2 = Tz_2 = 4u_1 - u_2 - 2u_3$. From Theorem 4.6.6, we may choose the ordered $\mathbb{Q}$-basis for J to be

$$\begin{aligned} \gamma' &= (z_1, z_2, (\lambda - 1)z_2) = (z_1, z_2, \lambda z_2 - z_2) \\ &= (-2u_1 + u_3, u_1, 3u_1 - u_2 - 2u_3). \end{aligned}$$

We may choose P to be

$$\begin{pmatrix} -2 & 1 & 3 \\ 0 & 0 & -1 \\ 1 & 0 & -2 \end{pmatrix}$$

since it is the base change matrix from γ' to (u_1, u_2, u_3). ◇

Example 4.6.10. Let F be a field and let

$$A = \begin{pmatrix} 0 & 0 & -1 \\ 1 & 0 & 1 \\ 0 & 1 & 1 \end{pmatrix} \in M_3(F).$$

Find a Jordan form J for A and find P such that $J = P^{-1}AP$.

Jordan forms are usually preferable to rational forms.

Solution. Notice that A is the companion matrix of

$$f(\lambda) = \lambda^3 - \lambda^2 - \lambda + 1 = (\lambda + 1)(\lambda - 1)^2.$$

Hence $f(\lambda)$ is the sole invariant factors of $\operatorname{Coker}(\lambda I - A)$. It follows that the elementary divisors of $\operatorname{Coker}(\lambda I - A)$ are $\lambda + 1$ and $(\lambda - 1)^2$. Thus the matrix

$$J = \left(\begin{array}{c|cc} -1 & & \\ \hline & 1 & \\ & 1 & 1 \end{array} \right)$$

is a Jordan form for A over F.

If one wants to find a basis for a rational form of A, it is easier to use the method in Example 4.5.14, which will incidentally give a basis for a Jordan form (if it exists) as well. However, in this example we only need to find a basis for J. We will propose a different but (usually) simpler method.

Let $T: F^3 \to F^3$ be the F-linear endomorphism represented by A with respect to the standard ordered basis $\alpha = (e_1, e_2, e_3)$. If we can find an ordered basis $\beta = (v_1, v_2, v_3)$ with respect to which J represents T, we should have

$$Tv_1 = -v_1, \quad Tv_2 = v_2 + v_3 \quad \text{and} \quad Tv_3 = v_3.$$

Hence v_1 and v_3 are eigenvectors associated with -1 and 1 respectively. We may find

$$v_1 \in \operatorname{Ker}(A + I) = \operatorname{Ker} \begin{pmatrix} 1 & 0 & -1 \\ 1 & 1 & 1 \\ 0 & 1 & 2 \end{pmatrix} = \langle e_1 - 2e_2 + e_3 \rangle;$$

$$v_3 \in \mathrm{Ker}\,(A - I) = \mathrm{Ker} \begin{pmatrix} -1 & 0 & -1 \\ 1 & -1 & 1 \\ 0 & 1 & 0 \end{pmatrix} = \langle -e_1 + e_3 \rangle.$$

We may choose $v_1 = e_1 - 2e_2 + e_3$ and $v_3 = -e_1 + e_3$. Since $(T - \mathbf{1}_{F^3})v_2 = v_3$, we need to find a special solution for the equation

$$\begin{pmatrix} -1 & 0 & -1 \\ 1 & -1 & 1 \\ 0 & 1 & 0 \end{pmatrix} \begin{pmatrix} x \\ y \\ z \end{pmatrix} = \begin{pmatrix} -1 \\ 0 \\ 1 \end{pmatrix}.$$

We have that $(x, y, z) = (0, 1, 1)$ is such a solution (the choice is not unique). We may choose $v_2 = e_2 + e_3$. Note that v_2 is a vector in $\mathrm{Ker}\,(T - \mathbf{1}_{F^3})^2$ but not in $\mathrm{Ker}\,(T - \mathbf{1}_{F^3})$. Hence the base change matrix from β to α

$$P = \begin{pmatrix} 1 & 0 & -1 \\ -2 & 1 & 0 \\ 1 & 1 & 1 \end{pmatrix}$$

is one of the many choices of P such that $J = P^{-1}AP$. In Exercise 6 the reader is asked to find other suitable bases for the Jordan form J.   ◇

Example 4.6.11. Find a Jordan form J for the matrix

$$A = \begin{pmatrix} 1 & 0 & 0 & 0 \\ 9 & 4 & -5 & 1 \\ 8 & 3 & -4 & 1 \\ 1 & 2 & -2 & 1 \end{pmatrix}$$

over $\mathbb{Q}$. Find an invertible matrix P over $\mathbb{Q}$ such that $J = P^{-1}AP$.

Solution. We will skip the details of finding the normal form for $\lambda I - A$. We will simply say that the elementary divisors of $\mathrm{Coker}\,(\lambda I - A)$ are $(\lambda - 1)^3$ and $\lambda + 1$. Hence we may choose

$$J = \left(\begin{array}{ccc|c} 1 & & & \\ 1 & 1 & & \\ & 1 & 1 & \\ \hline & & & -1 \end{array} \right)$$

to be a Jordan form for A. We will proceed to find P.

Let $T: \mathbb{Q}^4 \to \mathbb{Q}^4$ be the $\mathbb{Q}$-linear transformation whose matrix with respect to the standard ordered basis α is A. We need to find an ordered basis $\beta = (v_1, v_2, v_3, v_4)$ so that J is the matrix of T with respect to β. By observing J we may choose

$$v_1 \in \text{Ker} \, (T - \mathbf{1}_{\mathbb{Q}})^3 \setminus \text{Ker} \, (T - \mathbf{1}_{\mathbb{Q}})^2;$$
$$v_2 = (T - \mathbf{1}_{\mathbb{Q}})(v_1);$$
$$v_3 = (T - \mathbf{1}_{\mathbb{Q}})(v_2);$$
$$v_4 \in \text{Ker} \, (T + \mathbf{1}_{\mathbb{Q}}).$$

The matrix of $T - \mathbf{1}_{\mathbb{Q}}$ with respect to α is

$$A - I = \begin{pmatrix} 1 & 0 & 0 & 0 \\ 9 & 4 & -5 & 1 \\ 8 & 3 & -4 & 1 \\ 1 & 2 & -2 & 1 \end{pmatrix} - \begin{pmatrix} 1 & 0 & 0 & 0 \\ 0 & 1 & 0 & 0 \\ 0 & 0 & 1 & 0 \\ 0 & 0 & 0 & 1 \end{pmatrix} = \begin{pmatrix} 0 & 0 & 0 & 0 \\ 9 & 3 & -5 & 1 \\ 8 & 3 & -5 & 1 \\ 1 & 2 & -2 & 0 \end{pmatrix}.$$

Consequently we have

$$(A - I)^2 = \begin{pmatrix} 0 & 0 & 0 & 0 \\ -12 & -4 & 8 & -2 \\ -12 & -4 & 8 & -2 \\ 2 & 0 & 0 & 0 \end{pmatrix}; \quad (A - I)^3 = \begin{pmatrix} 0 & 0 & 0 & 0 \\ -52 & -16 & 32 & -8 \\ -52 & -16 & 32 & -8 \\ 0 & 0 & 0 & 0 \end{pmatrix}.$$

To find v_1, we need to find the null space of $(A - I)^3$. We have that

$$\text{Ker} \, (T - \mathbf{1}_{\mathbb{Q}})^3 = \langle (1, 0, 0, -6.5), \, (0, 1, 0, -2), \, (0, 0, 1, 4) \rangle.$$

Since $(1, 0, 0, -6.5) \notin \text{Ker} \, (T - \mathbf{1}_{\mathbb{Q}})^2$, we may choose

$$v_1 = (1, 0, 0, -6.5);$$
$$v_2 = (T - \mathbf{1}_{\mathbb{Q}})(v_1) = (0, 2.5, 1.5, 1);$$
$$v_3 = (T - \mathbf{1}_{\mathbb{Q}})(v_2) = (0, 1, 1, 2).$$

The matrix of $T + \mathbf{1}_{\mathbb{Q}}$ with respect to α is

$$A + I = \begin{pmatrix} 1 & 0 & 0 & 0 \\ 9 & 4 & -5 & 1 \\ 8 & 3 & -4 & 1 \\ 1 & 2 & -2 & 1 \end{pmatrix} + \begin{pmatrix} 1 & 0 & 0 & 0 \\ 0 & 1 & 0 & 0 \\ 0 & 0 & 1 & 0 \\ 0 & 0 & 0 & 1 \end{pmatrix} = \begin{pmatrix} 2 & 0 & 0 & 0 \\ 9 & 5 & -5 & 1 \\ 8 & 3 & -3 & 1 \\ 1 & 2 & -2 & 2 \end{pmatrix}.$$

We have that

$$\mathrm{Ker}\,(T + \mathbf{1}_{\mathbb{Q}}) = \langle (0,\, 1,\, 1,\, 0) \rangle.$$

We may choose $v_4 = (0,\, 1,\, 1,\, 0)$.

With respect to the ordered basis $\beta = (v_1,\, v_2,\, v_3,\, v_4)$, the matrix of the linear endomorphism T is J. Hence we may choose

$$P = \begin{pmatrix} 1 & 0 & 0 & 0 \\ 0 & 2.5 & 1 & 1 \\ 0 & 1.5 & 1 & 1 \\ -6.5 & 1 & 2 & 0 \end{pmatrix},$$

the base change matrix from β to α.   ◇

Uniqueness of the canonical forms

Finally we will discuss how to use the canonical forms (rational or Jordan) to classify similar matrices.

Theorem 4.6.12. *Let V be a finite dimensional vector space over the field F. Suppose T is an F-linear endomorphism on V. The rational form of a linear endomorphism T or of a matrix A is unique. The Jordan form of T or of A (if it exists) is unique up to different ordering of the Jordan blocks.*

Proof. Suppose given a rational form of T

$$B = \begin{pmatrix} B_1 & & & \\ & B_2 & & \\ & & \ddots & \\ & & & B_s \end{pmatrix}$$

where B_i is the companion matrix of $d_i(\lambda)$ and $d_i \mid d_{i+1}$ for each i. Let $\deg d_i(\lambda) = n_i$. We may find an ordered basis

$$\beta = (u_{11},\, u_{12},\, \ldots,\, u_{1n_1};\, u_{21},\, u_{22},\, \ldots,\, u_{2n_2};\, \ldots;\, u_{s1},\, u_{s2},\, \ldots,\, u_{sn_s})$$

such that B is the matrix of T with respect to β. Then

$$V_i = \langle u_{i1},\, u_{i2},\, \ldots,\, u_{in_i} \rangle$$

is T-invariant for each i and B_i is the matrix for T_{V_i} with respect to the ordered basis $(u_{i1}, u_{i2}, \ldots, u_{in_1})$. From Exercise 12, §4.5, we have that

$$V = \bigoplus_{i=1}^{s} V_i \cong \bigoplus_{i=1}^{s} \frac{F[\lambda]}{(d_i(\lambda))}$$

via T. Since $d_i \mid d_{i+1}$ for all i, the $d_i(\lambda)$'s are the invariant factors of V via T. From the invariance theorem (Theorem 4.4.21), the invariant factors are completely determined by T. Thus the rational form of T is unique.

Suppose T has a Jordan form

$$J = \begin{pmatrix} J_1 & & & \\ & J_2 & & \\ & & \ddots & \\ & & & J_t \end{pmatrix}$$

where J_i is the Jordan block with respect to r_i of size e_i. Let

$$\gamma = (v_{11}, v_{12}, \ldots, v_{1e_1}; v_{21}, v_{22}, \ldots, v_{2e_2}; \ldots; v_{t1}, v_{t2}, \ldots, v_{te_t})$$

be an ordered basis with respect to which J is the matrix representing T. Then

$$W_i = \langle v_{i1}, v_{i2}, \ldots, v_{ie_i} \rangle$$

is T-invariant for each i and J_i is the matrix of T_{W_i} with respect to the ordered basis $(v_{i1}, v_{i2}, \ldots, v_{ie_i})$. We leave it as an easy exercise (see Exercise 3) to verify that

$$V = \bigoplus_{i=1}^{t} W_i \cong \bigoplus_{i=1}^{t} \frac{F[\lambda]}{((\lambda - r_i)^{e_i})}$$

via T. The $(\lambda - r_i)^{e_i}$'s are the elementary divisors of V via T. They are completely determined by T from the invariance theorem again. Hence the Jordan form of T is unique up to different ordering of the Jordan blocks. $\square$

To summarize:

- The invariant factors determine the rational canonical form.

- The elementary divisors determine the invariant factors, and hence determine the rational canonical form.

- If the elementary divisors are all powers of linear polynomials, they determine the Jordan canonical form.

Corollary 4.6.13. *Two rational forms are similar if and only if they have the same companion matrices (counted with multiplicity). Two Jordan forms are similar if and only if they have the same Jordan blocks (counted with multiplicity). Two square matrices over a filed F are similar if and only if they have similar canonical forms.*

Example 4.6.14. Determine whether the following matrices are similar:

$$
A = \begin{pmatrix} 0 & 0 & 4 & & & \\ 1 & 0 & -8 & & & \\ 0 & 1 & 5 & & & \\ & & & 0 & 0 & 2 \\ & & & 1 & 0 & -5 \\ & & & 0 & 1 & 4 \end{pmatrix} \quad \text{and} \quad B = \begin{pmatrix} 0 & -2 & & & & \\ 1 & 3 & & & & \\ & & 0 & 0 & 0 & -4 \\ & & 1 & 0 & 0 & 12 \\ & & 0 & 1 & 0 & -13 \\ & & 0 & 0 & 1 & 6 \end{pmatrix}.
$$

Solution. The blocks in A and B are all companion matrices. It is easy to see that

$$
\begin{aligned}
\operatorname{Coker}(\lambda I - A) &\cong \frac{F[\lambda]}{(\lambda^3 - 5\lambda^2 + 8\lambda - 4)} \oplus \frac{F[\lambda]}{(\lambda^3 - 4\lambda^2 + 5\lambda - 2)} \\
&= \frac{F[\lambda]}{((\lambda - 1)(\lambda - 2)^2)} \oplus \frac{F[\lambda]}{((\lambda - 1)^2(\lambda - 2))}; \\
\operatorname{Coker}(\lambda I - B) &\cong \frac{F[\lambda]}{(\lambda^2 - 3\lambda + 2)} \oplus \frac{F[\lambda]}{(\lambda^4 - 6\lambda^3 + 13\lambda^2 - 12\lambda + 4)} \\
&\cong \frac{F[\lambda]}{((\lambda - 1)(\lambda - 2))} \oplus \frac{F[\lambda]}{((\lambda - 1)^2(\lambda - 2)^2)}.
\end{aligned}
$$

The elementary divisors of both $\operatorname{Coker}(\lambda I - A)$ and $\operatorname{Coker}(\lambda I - B)$ are

$$
\lambda - 1, \quad \lambda - 2, \quad (\lambda - 1)^2 \quad \text{and} \quad (\lambda - 2)^2.
$$

Thus A and B are similar to each other. In fact, B is the rational canonical form of A.   ◇

Definition 4.6.15. We say a square matrix over the field F is **diagonalizable** if it is similar to a diagonal matrix over F.

If A is diagonalizable and A represents the linear endomorphism T on V, then V contains a basis consisting of eigenvectors. On the other hand, a diagonal matrix is also a Jordan form all of whose Jordan blocks are of size 1. The elementary divisors of A are all linear.

Corollary 4.6.16. *The following statements are equivalent for a square matrix A over a field F.*

(a) *The matrix A is diagonalizable over F.*

(b) *The matrix A has a Jordan form and the Jordan form is a diagonal matrix.*

(c) *The minimal polynomial of A is a product of distinct monic polynomials of degree 1 over F.*

(d) *Let the degree of the minimal polynomial of A be m. The minimal polynomial has m distinct zeros in F.*

Corollary 4.6.17. *If a square matrix of size n has n distinct eigenvalues over a field F, it is diagonalizable over F.*

We leave the proofs of these two corollaries as exercises. See Exercise 7.

Let's look at some applications of the canonical forms.

Definition 4.6.18. We say a linear endomorphism T on V is a **projection** if $T^2 = T$. We say T is **nilpotent** if $T^m = 0$ for some m.

Example 4.6.19. Classify all projections on finite dimensional vector spaces.

Solution. Suppose T is a projection on a finite dimensional vector space V over the field F. Let $g(\lambda) = \lambda^2 - \lambda$. Then $g(T) = 0$. The minimal polynomial $m(\lambda)$ of T is a divisor of $g(\lambda)$ by the Cayley-Hamilton Theorem (Theorem 4.5.8). Hence $m(\lambda) = \lambda$, $\lambda - 1$ or $\lambda(\lambda - 1)$.

    Case 1. Assume that $m(\lambda) = \lambda$. Then $m(T) = T = 0$ by Theorem 4.5.8(c). Thus T is the trivial endomorphism.

    Case 2. Assume that $m(\lambda) = \lambda - 1$. We have $m(T) = T - 1_V = 0$. Thus T is the identity map on V.

Case 3. Assume that $m(\lambda) = \lambda(\lambda - 1)$. In this case the elementary divisors of V via T are a series of λ and a series of $\lambda - 1$. Hence the Jordan form of T is of the form

$$\begin{pmatrix} 1 & & & & \\ & \ddots & & & \\ & & 1 & & \\ & & & 0 & \\ & & & & 0 \end{pmatrix}.$$

This shows that for an appropriate basis for V, say, $(u_1, u_2, \ldots, u_n)$, there exists r with $0 < r < n$ such that $T\left(\sum_{i=1}^{n} a_i u_i\right) = \sum_{i=1}^{r} a_i u_i$. This fits our preconceived idea about *projections*.   ◇

Example 4.6.20. Let N be a nilpotent matrix of size 5 over $\mathbb{R}$. Find all possible Jordan forms for N. What if the nilpotent matrices are over $\mathbb{C}$? Does the scalar field matter?

Solution. If N is nilpotent, there exists m such that $N^m = 0$. By the Cayley-Hamilton theorem, the minimal polynomial $m(\lambda)$ of N is a divisor of λ^m. Hence $m(\lambda) = \lambda^t$ for some positive integer t. The elementary divisors of $\mathrm{Coker}\,(\lambda I - N)$ are a series of powers of λ whose powers are no greater than t. Thus the Jordan form of N consists of Jordan blocks with respect to the eigenvalue 0 of various sizes. When N is of size 5, the Jordan form of N is one of the following matrices

$$J(0, 5), \quad \left(\begin{array}{c|c} 0 & \\ \hline & J(0, 4) \end{array}\right), \quad \left(\begin{array}{c|c} J(0, 2) & \\ \hline & J(0, 3) \end{array}\right),$$

$$\left(\begin{array}{c|c|c} 0 & & \\ \hline & 0 & \\ \hline & & J(0, 3) \end{array}\right), \quad \left(\begin{array}{c|c|c} 0 & & \\ \hline & J(0, 2) & \\ \hline & & J(0, 2) \end{array}\right),$$

$$\left(\begin{array}{c|c|c|c} 0 & & & \\ \hline & 0 & & \\ \hline & & 0 & \\ \hline & & & J(0, 2) \end{array}\right) \quad \text{and} \quad \left(\begin{array}{c|c|c|c|c} 0 & & & & \\ \hline & 0 & & & \\ \hline & & 0 & & \\ \hline & & & 0 & \\ \hline & & & & 0 \end{array}\right),$$

where $J(0, e)$ denotes the Jordan block with respect to 0 of size e. The discussion above is totally independent of the scalar field. The solution does not change no matter if we consider the nilpotent matrix over $\mathbb{R}$, over $\mathbb{C}$ or over any field F.   ◇

Example 4.6.21. Let T be a linear endomorphism on the n-dimensional F-vector space V with n distinct eigenvalues. Show that there exists a vector $v \in V$ such that $v, Tv, \ldots, T^{n-1}v$ form a basis for V over F.

Solution. Let $r_1, r_2, \ldots, r_n$ be the n-distinct eigenvalues of T. This implies that elementary divisors of V via T are $\lambda - r_1, \lambda - r_2, \ldots, \lambda - r_n$. In other words, $\prod_{i=1}^{n}(\lambda - r_i)$ is the sole invariant factor of V via T. Hence, there exists $v \in V$ such that

$$V = F[\lambda]v \cong \frac{F[\lambda]}{\left(\prod_{i=1}^{n}(\lambda - r_i)\right)}$$

is a cyclic module. From Lemma 4.5.11 we have that $v, \lambda v, \ldots, \lambda^{n-1}v$ form a basis for V over F. This translates to the fact that $v, Tv, \ldots, T^{n-1}v$ form a basis for V over F.

However, for someone who has no knowledge on module theory, the argument above will not be understood. We provide an more elementary argument just for fun.

An Alternative Method. Let v_i be an eigenvector with respect to the eigenvalue r_i for $i = 1, 2, \ldots, n$. We claim that $v = v_1 + v_2 + \cdots + v_n$ is what we seek. Note that the coordinate of $T^i v$ with respect to the ordered basis $(v_1, v_2, \ldots, v_n)$ is $(r_1^i, r_2^i, \ldots, r_n^i)$ for $i = 0, 1, \ldots, n - 1$. From Proposition 4.1.1, we have that $v, Tv, \ldots, T^{n-1}v$ form a basis for V over F if the Vandermonde matrix

(4.6.4)
$$\begin{pmatrix} 1 & r_1 & r_1^2 & \cdots & r_1^{n-1} \\ 1 & r_2 & r_2^2 & \cdots & r_2^{n-1} \\ \cdots\cdots\cdots\cdots\cdots\cdots \\ \cdots\cdots\cdots\cdots\cdots\cdots \\ 1 & r_n & r_n^2 & \cdots & r_n^{n-1} \end{pmatrix}$$

is invertible. This is indeed so since the determinant of the Vandermonde matrix in (4.6.4) is $\displaystyle\prod_{1 \leq i < j \leq n}(r_j - r_i) \neq 0$ by Exercise 8, §3.1.   ◇

Exercises 4.6

Throughout these exercises F denotes a field, V denotes a finite dimensional F-vector space and λ denotes an indeterminate over F.

1. Prove Lemma 4.6.2.

2. Let $f \in F[\lambda]$ with $\deg f = n$. Let $g_0, g_1, \ldots, g_{n-1} \in F[\lambda]$ be such that $\deg g_i = i$ for each i. Show that $\overline{g}_0, \overline{g}_1, \ldots, \overline{g}_{n-1}$ form a basis for $V = F[\lambda]/(f)$ over F.

3. Let $r \in F$ and let V be an e-dimensional vector space over F. If the Jordan block with respect to r of size e in (4.6.2) represents an F-linear endomorphism $T \colon V \to V$, show that via T, V is a cyclic $F[\lambda]$-module whose order ideal is $((\lambda - r)^e)$.

4. Show that both the minimal and the characteristic polynomials of the Jordan block J in (4.6.2) are $(\lambda - r)^e$ by working directly on $\lambda I - J$.

   Show that the characteristic polynomial of the matrix J in (4.6.3) is $\prod_{i=1}^{t} (\lambda - r_i)^{e_i}$.

5. Let J be a Jordan block with entries in F. Show that J^{tr} is similar to J over F.

6. Following the notation in Example 4.6.10, find two other invertible matrices P_1 and P_2 such that $J = P_1^{-1}AP_1 = P_2^{-1}AP_2$.

7. Verify Corollary 4.6.16 and Corollary 4.6.17.

8. Let T be an F-linear endomorphism on V. Show that T is diagonalizable over F if and only if V is a direct sum of eigenspaces.

9. Find the Jordan form for each of the following matrices over $\mathbb{C}$.

   (a) $\begin{pmatrix} 0 & 0 & 1 \\ 1 & 0 & -3 \\ 0 & 1 & 3 \end{pmatrix}$ (b) $\begin{pmatrix} 0 & 0 & -1 \\ 1 & 0 & 1 \\ 0 & 1 & 1 \end{pmatrix}$ (c) $\begin{pmatrix} 0 & 1 & & & \\ 1 & 0 & & & \\ \hline & & 0 & 0 & 1 \\ & & 1 & 0 & 0 \\ & & 0 & 1 & 0 \end{pmatrix}$.

10. Find the rational form for

$$\begin{pmatrix} 2 & & & & & & & \\ & 0 & & & & & & \\ & 1 & 0 & & & & & \\ & & & 0 & & & & \\ & & & 1 & 0 & & & \\ & & & & 1 & 0 & & \\ & & & & & & 1 & \\ & & & & & & 1 & 1 \end{pmatrix} \in M_8(\mathbb{Q}).$$

11. Determine whether the matrix

$$\begin{pmatrix} 1 & 0 & 0 & 0 \\ 0 & 1 & 0 & 0 \\ -2 & -2 & 0 & 1 \\ -2 & 0 & -1 & -2 \end{pmatrix}$$

possesses a Jordan form over $\mathbb{Q}$. Find the Jordan form if it exists.
(*Cf.* Exercise 9, §4.5.)

12. Let

$$A = \begin{pmatrix} 4 & -3 & 9 & 0 & -3 & 0 \\ 0 & 1 & 0 & 0 & 1 & 0 \\ -1 & 1 & -2 & 0 & 1 & 0 \\ 1 & 0 & 3 & 1 & -1 & 0 \\ 0 & 0 & 0 & 0 & 1 & 0 \\ 0 & 0 & 1 & 1 & 0 & 1 \end{pmatrix} \in M_6(\mathbb{Q}).$$

Determine whether the matrix A has a Jordan form over $\mathbb{Q}$. Find the
Jordan form J of the matrix A if it exists. Find an invertible matrix
P such that $J = P^{-1}AP$ when J exists. (*Cf.* Exercise 10, §4.5.)

13. Let

$$A = \begin{pmatrix} -1 & 0 & -3 & -3 & 0 & 3 \\ 0 & -1 & 0 & 0 & 1 & 0 \\ 0 & 0 & 0 & 1 & 0 & -1 \\ 1 & 0 & 3 & -1 & -1 & 0 \\ 0 & 0 & 0 & 0 & -1 & 0 \\ -1 & 0 & -2 & 1 & 1 & 0 \end{pmatrix} \in M_6(\mathbb{Q}).$$

Determine whether A has a Jordan form over $\mathbb{Q}$. Determine the Jordan form J of the matrix A if it exists. Find an invertible matrix P such that $J = P^{-1}AP$ if J exists. (*Cf.* Exercise 11, §4.5.)

14. Determine whether the two matrices

$$
\begin{pmatrix}
0 & 1 & 0 & \cdots & 0 \\
0 & 0 & 1 & \cdots & 0 \\
\multicolumn{5}{c}{\dotfill} \\
\multicolumn{5}{c}{\dotfill} \\
\multicolumn{3}{c}{\dotfill} & 0 & 1 \\
1 & 0 & \multicolumn{3}{c}{\dotfill} & 0
\end{pmatrix}
\quad \text{and} \quad
\begin{pmatrix}
1 & 1 & 0 & \cdots & 0 \\
0 & 1 & 1 & 0 & \cdots \\
\multicolumn{5}{c}{\dotfill} \\
\multicolumn{5}{c}{\dotfill} \\
\multicolumn{3}{c}{\dotfill} & 1 & 1 \\
\multicolumn{3}{c}{\dotfill} & 0 & 1
\end{pmatrix}
$$

are similar over $\mathbb{Z}_p$.

15. Let V be a finite dimensional $\mathbb{C}$-vector space.

   (a) Show that every diagonalizable operator is semisimple.

   (b) **(Jordan-Chevalley decomposition.)** Show that every linear endomorphism T on V may be decomposed as $T = S + N$ where S is semisimple and N is nilpotent, and the decomposition is unique. Show that $SN = NS$.

16. Let $\{x_{ij} : i, j = 1, \ldots, n\}$ be n^2 indeterminates over the field F. Show that the square matrix $A = \left(x_{ij}\right)_{n \times n}$ is diagonalizable over an algebraically closed field E containing $F(x_{ij} : i, j = 1, \ldots, n)$. (Hint: Show that the characteristic polynomial of A has n simple zeros in E.)

CHAPTER 5

A Brief Introduction to the
Tensor Product

My original plan for this book was to stop at Chapter 4. However, some of my more adventurous students encountered tensor products and needed help. Since the tensor product is indeed somewhat confusing and intriguing to beginners, I decided to add one bonus chapter for my students and readers who are also interested in this topic.

We will first give a definition of the tensor product for finite dimensional vector spaces. The first approach is concrete and serves as a motivation for the general case. Then we will give the second definition of the tensor product which is abstract but works for modules in general. The tensor product is directly related to bilinear maps. Hence, this chapter will start with the discussion of bilinear maps.

5.1   Bilinear maps and multilinear maps

In this section we will generalize the concept of linear maps and linear functionals to bilinear maps and bilinear forms.

Throughout this section R denotes a ring and F a field.

Bilinear maps and multilinear maps

Throughout this subsection, M, N and W denote modules over R.

Definition 5.1.1. A function B from $M \times N$ to W is called an R-**bilinear map** or a bilinear map over R if B satisfies the following conditions

(i) $B(a_1 m_1 + a_2 m_2, \, n) = a_1 B(m_1, \, n) + a_2 B(m_2, \, n)$, and

(ii) $B(m, \, a_1 n_1 + a_2 n_2) = a_1 B(m, \, n_1) + a_2 B(m, \, n_2)$

for all $a_1, a_2 \in R$, $m, m_1, m_2 \in M$ and $n, n_1, n_2 \in N$.

When $W = R$, the R-bilinear map is also called an R-**bilinear form** or a bilinear form over R.

A bilinear map is linear at one coordinate while the other coordinate is fixed. To be precise, the maps $B(m, \, -)$ and $B(-, \, n)$ are linear for any $m \in M$ and $n \in N$.

Example 5.1.2. Let a be an element in R. The map $R \times R \to R$ sending (x, y) to axy is R-bilinear.

Example 5.1.3. The map

$$B\big((x, y), (z, w)\big) = (xz + yw, \, xz + 2xw + 3yz + 4yw), \qquad x, y, z, w \in R$$

is an R-bilinear map from $R^2 \times R^2$ to R^2.

Example 5.1.4. Let V be a vector space over $F = \mathbb{R}$ or $\mathbb{C}$. An *inner product*, denoted $\langle -, \, - \rangle$, on V is an F-bilinear form on $V \times V$ such that

(i) $\langle v, u \rangle = \overline{\langle u, v \rangle}$,

(ii) $\langle v, v \rangle \geq 0$, and

(iii) if $\langle v, v \rangle = 0$ then $v = 0$

for all u and $v \in V$.

Note that from (i) we have that $\langle v, v \rangle \in \mathbb{R}$ even when $F = \mathbb{C}$. Thus the condition in (ii) always makes sense.

Remember that M^* denotes the dual module $\text{Hom}_R(M, R)$ (see Exercises 7 and 15, §2.1).

Example 5.1.5. Define

$$\begin{array}{rcl} B: & M \times M^* & \longrightarrow & R \\ & (m, f) & \longmapsto & f(m) \end{array} .$$

Let $a, b \in R$, $m, n \in M$ and $f, g \in M^*$. Then

$$B(am + bn, f) = f(am + bn) = af(m) + bf(n) = aB(m, f) + bB(n, f)$$
$$B(m, af + bg) = (af + bg)(m) = af(m) + bg(m) = aB(m, f) + bB(m, g)$$

since f is linear. Thus B is a bilinear form over R.

Example 5.1.6. Let $f \in M^*$ and $g \in N^*$. Define

$$\begin{array}{rcl} B: & M \times N & \longrightarrow & R \\ & (m, n) & \longmapsto & f(m)g(n) \end{array} .$$

Let $a, a' \in R$, $m, m' \in M$ and $n, n' \in N$. Then

$$B(am + a'm', n) = f(am + a'm')g(n) = \big(af(m) + a'f(m')\big)g(n)$$
$$= af(m)g(n) + a'f(m')g(n) = aB(m, n) + a'B(m', n);$$
$$B(m, an + a'n') = f(m)g(an + a'n') = f(m)\big(ag(n) + a'g(n')\big)$$
$$= af(m)g(n) + a'f(m)g(n') = aB(m, n) + a'B(m, n').$$

Thus B is a bilinear form over R.

See Exercise 2 for a natural generalization of this example.

We will use the notation $\text{Bil}_R(M \times N, W)$ for the set of all bilinear maps from $M \times N$ to W over R. For $B, B' \in \text{Bil}_R(M \times N, W)$ and $a \in R$, there is a natural addition and a natural scalar multiplication defined as follows:

(5.1.1)
$$(B + B')(m, n) = B(m, n) + B'(m, n), \quad \text{and}$$
$$(aB)(m, n) = aB(m, n)$$

for $(m, n) \in M \times N$.

Lemma 5.1.7. *Equipped with* (5.1.1) *the set* $\mathrm{Bil}_R(M \times N, W)$ *becomes an R-module.*

We leave this lemma as an easy exercise. See Exercise 3.

The concept of bilinear maps may be naturally generalized to the following concept.

Definition 5.1.8. Let k be a positive integer. Let M_1, M_2, ..., M_k and W be modules over R. A map from $M_1 \times M_2 \times \cdots \times M_k$ to W is called a **multilinear map**, or more specifically a k-**linear map** over R, if for each i, the map $f(m_1, \ldots, m_i, \ldots, m_n)$ is a linear map of m_i while all other variables but m_i are held constant. When $W = R$, the linear map f is called a **multilinear form** or more specifically a k-**linear form** over R.

Example 5.1.9. The determinant of a square matrix in $M_n(R)$ is an multilinear form of the columns of the matrix over R.

Bilinear forms

Throughout this subsection we will assume M and N are free R-modules of finite rank. We will also assume that $\beta = (u_1, u_2, \ldots, u_s)$ is an ordered basis for M over R and $\gamma = (v_1, v_2, \ldots, v_t)$ is an ordered basis for N over R.

Find the dual bases $\beta' = (f_1, f_2, \ldots, f_s)$ of β and $\gamma' = (g_1, g_2, \ldots, g_t)$ of γ. (See Exercise 15, §2.1 for the definition of dual bases.) For each $1 \leq i \leq s$ and $1 \leq j \leq t$, define B_{ij} to be the bilinear form sending (m, n) to $f_i(m)g_j(n)$ for $m \in M$ and $n \in N$ (see Example 5.1.6). Note that

$$(5.1.2) \qquad B_{ij}(u_k, v_\ell) = \begin{cases} 1, & \text{if } k = i \text{ and } \ell = j; \\ 0, & \text{otherwise.} \end{cases}$$

Furthermore, we have the following relations for all i and j:

$$(5.1.3) \qquad B_{ij}\left(\sum_k a_k u_k, \sum_\ell b_\ell v_\ell\right) = \sum_{k,\ell} a_k b_\ell B_{ij}(u_k, v_\ell) = a_i b_j;$$

$$(5.1.4) \qquad \left(\sum_{k,\ell} a_{k\ell} B_{k\ell}\right)(u_i, v_j) = \sum_{k,\ell} a_{k\ell} B_{k\ell}(u_i, v_j) = a_{ij}.$$

Proposition 5.1.10. *Let M and N be free R-modules of rank s and t respectively. Let $\beta = (u_1, u_2, \ldots, u_s)$ and $\gamma = (v_1, v_2, \ldots, v_t)$ be ordered R-bases for M and N respectively. Let B_{ij}, $1 \leq i \leq s$ and $1 \leq j \leq t$, be given as in (5.1.2). Then*

$$\mathscr{B} = \big\{ B_{ij} \in \mathrm{Bil}_R(M \times N, R) : 1 \leq i \leq s, \ 1 \leq j \leq t \big\}$$

is a basis for $\mathrm{Bil}_R(M \times N, R)$ over R. It follows that

$$\mathrm{rk}_R \mathrm{Bil}_R(M \times N, R) = st = \mathrm{rk}_R M \cdot \mathrm{rk}_R N.$$

Proof. Suppose $\sum_{k, \ell} a_{k\ell} B_{k\ell} = 0$, where $a_{k\ell} \in R$. By (5.1.4) we have that

$$0 = \left(\sum_{k, \ell} a_{k\ell} B_{k\ell} \right) (u_i, v_j) = a_{ij}$$

for $1 \leq i \leq s$ and $1 \leq j \leq t$. Thus $\mathscr{B}$ is linearly independent over R.

Next we show that $\mathscr{B}$ generates $\mathrm{Bil}_R(M \times N, R)$ over R. We are done if we can prove the claim that

(5.1.5) $$B = \sum_{k, \ell} a_{k\ell} B_{k\ell}, \qquad \text{where } a_{k\ell} = B(u_k, v_\ell)$$

for any bilinear form B on $M \times N$.

Let $m = \sum_i b_i u_i \in M$ and $n = \sum_j c_j v_j \in N$. Then

$$\left(\sum_{k, \ell} a_{k\ell} B_{k\ell} \right) (m, n) = \left(\sum_{k, \ell} a_{k\ell} B_{k\ell} \right) \left(\sum_{i=1}^{s} b_i u_i, \ \sum_{j=1}^{t} c_j v_j \right)$$

$$= \sum_{k, \ell} a_{k\ell} B_{k\ell} \left(\sum_{i=1}^{s} b_i u_i, \ \sum_{j=1}^{t} c_j v_j \right) = \sum_{k, \ell} a_{k\ell} b_k c_\ell$$

by (5.1.3), while

$$B(m, n) = B \left(\sum_{k=1}^{m} b_k u_k, \ \sum_{\ell=1}^{n} c_\ell v_\ell \right)$$

(5.1.6)

$$= \sum_{k, \ell} b_k c_\ell B(u_k, \ v_\ell) = \sum_{k, \ell} b_k c_\ell a_{k\ell}.$$

Hence the claim in (5.1.5) is valid. $\qquad \square$

Let $m \in M$. Remember that in §2.3, we used m_β^{row} to denote the coordinates of m with respect to β as a row vector, and m_β^{col} to denote the same as a column vector. Here we will follow these notations as well.

Let B be a bilinear form on $M \times N$. Let

$$(5.1.7) \qquad A = \left(a_{ij}\right)_{s \times t}, \qquad \text{where } a_{ij} = B(u_i, v_j).$$

Let $m \in M$ and $n \in N$. The identity in (5.1.6) may be rewritten as

$$(5.1.8) \qquad B(m, n) = m_\beta^{\mathrm{row}} A n_\gamma^{\mathrm{col}}.$$

This is why we call the matrix A in (5.1.7) the **matrix of the bilinear form** B with respect to the ordered bases β and γ. Conversely, the matrix $A = \left(a_{ij}\right)_{s \times t}$ over R gives rise to a bilinear form $B = \sum_{k,\ell} a_{k\ell} B_{k\ell}$. Note that $B(u_i, v_j) = a_{ij}$ for all i and j by (5.1.4). Thus A is the matrix of B with respect to β and γ. We now have the following result.

Proposition 5.1.11. *Let M be a free R-module of rank s with β as an ordered basis. Let N be a free R-module of rank t with γ as an ordered basis. There is a one-to-one correspondence between $s \times t$ matrices and bilinear forms on $M \times N$ with respect to β and γ. More precisely, the bilinear form B corresponds with the matrix of B with respect to β and γ.*

Let $\beta_1 = (x_1, x_2, \ldots, x_s)$ and $\gamma_1 = (y_1, y_2, \ldots, y_t)$ also be respective ordered bases of M and N over R. Let P be the base change matrix from β_1 to β and let Q be the base change matrix from γ_1 to γ. Then $m_\beta^{\mathrm{col}} = P m_{\beta_1}^{\mathrm{col}}$ and $n_\gamma^{\mathrm{col}} = Q n_{\gamma_1}^{\mathrm{col}}$. Thus from (5.1.8) we have

$$B(m, n) = (m_{\beta_1}^{\mathrm{row}} P^{\mathrm{tr}}) A (Q n_{\gamma_1}^{\mathrm{col}}) = m_{\beta_1}^{\mathrm{row}} (P^{\mathrm{tr}} A Q) n_{\gamma_1}^{\mathrm{col}}.$$

We can now make the following conclusion.

Proposition 5.1.12. *Let M and N be free R-modules of finite rank. Let β, β_1 be ordered R-bases for M and let γ, γ_1 be ordered R-bases for N. If A is the matrix of the bilinear form B on $M \times N$ with respect to β and γ, then $P^{\mathrm{tr}} A Q$ is the matrix of B on $M \times N$ with respect to β_1 and γ_1, where P is the base change matrix from β_1 to β and Q is the base change matrix from γ_1 to γ.*

Exercises 5.1

In the following exercises, R denotes a ring and M, N, W, X are R-modules.

1. B is an R-bilinear map from $M \times N$ to W and L is a linear map from W to X. Show that LB is bilinear form $M \times N$ to X. Can you draw the same conclusion about k-linear maps?

2. Find $f_1, f_2, \ldots, f_n \in M^*$ and $g_1, g_2, \ldots, g_n \in N^*$. Define a function B from $M \times N$ to R by letting

$$B(m, n) = f_1(m)g_1(n) + f_2(m)g_2(n) + \cdots + f_n(m)g_n(n)$$

for $m \in M$ and $n \in N$. Show that B is a bilinear form over R.

3. Prove Lemma 5.1.7.

For the next two exercises, $\mathscr{P}_n$ denotes the subspace generated by monomials of degree $\leq n$ in $\mathbb{R}[x]$ where x is an indeterminate over $\mathbb{R}$.

4. In which of the following cases is B an $\mathbb{R}$-bilinear form on $\mathscr{P}_n \times \mathscr{P}_n$?

   (a) $B(f, g) = \int_0^1 f(x)g(x)\,dx$;

   (b) $B(f, g) = f(0) + g(0)$;

   (c) $B(f, g) = f(0)g(0)$;

   (d) $B(f, g) = f(0)g'(1)$.

5. In which of the following cases is B a bilinear map from $\mathscr{P}_n \times \mathscr{P}_n$ to $\mathbb{R}[x]$ over $\mathbb{R}$?

   (a) $B(f, g) = \int_0^x f(t)g(t)\,dt$;

   (b) $B(f, g) = f(x) + g(x)$;

   (c) $B(f, g) = f(x)g(x)$;

   (d) $B(f, g) = f(x)g(1)$;

   (e) $B(f, g) = f(0)g'(x)$;

(f) $B(f, g) = f(x)g'(x)$.

6. A bilinear form B on $M \times N$ is said to be **degenerate** if there exists a nonzero vector m_0 in M such that $B(m_0, -)$ is the trivial map on N, or there exists a nonzero vector n_0 in N such that $B(-, v_0)$ is the trivial map on M. Otherwise, we say the bilinear map B is **non-degenerate**.

   (a) Give an example of a non-trivial but degenerate bilinear form on $\mathbb{Q}^2 \times \mathbb{Q}^2$ over $\mathbb{Q}$.

   (b) Give an example of a non-degenerate bilinear form on $\mathbb{Q}^2 \times \mathbb{Q}^2$ over $\mathbb{Q}$.

7. Consider $V = \mathbb{Z}_2^2$ as a vector space over $\mathbb{Z}_2$. Let

$$A = \begin{pmatrix} 0 & 1 \\ 1 & 0 \end{pmatrix}$$

represent the bilinear form B on $V \times V$ with respect to the standard ordered basis of V (twice). Show that B is a non-trivial map such that $B(v, v) = 0$ for all $v \in V$.

8. Describe how to view $\mathrm{Tril}_R(M \times N \times W, X)$, the set of trilinear maps from $M \times N \times W$ to X over R, as an R-module.

Can you do the same for k-linear maps over R in general?

9. Let M, N and W be free R-modules of finite rank. Let

$$\beta = (u_1, u_2, \ldots, u_r), \quad \gamma = (v_1, v_2, \ldots, v_s), \quad \delta = (w_1, w_2, \ldots, w_t)$$

be ordered R-bases for M, N and W respectively. Let

$$\beta' = (f_1, f_2, \ldots, f_r), \quad \gamma' = (g_1, g_2, \ldots, g_s), \quad \delta' = (h_1, h_2, \ldots, h_t)$$

be their respective dual bases over R. Let T_{ijk} be the map from $M \times N \times W$ to R sending (m, n, p) to $f_i(m)g_j(n)h_k(p)$. Show that T_{ijk} is a trilinear form for each i, j and k. Show that

$$\mathscr{B} = \{T_{ijk} \in \mathrm{Tril}_R(M \times N \times W, R) : 1 \le i \le r, \ 1 \le j \le s, \ 1 \le k \le t\}$$

is an R-basis for $\mathrm{Tril}_R(M \times N \times W, R)$. Conclude that the rank of the module of the trilinear forms on $M \times N \times W$ is rst.

Can you speculate on the dimension of the general k-linear forms on free modules of finite rank?

There are plenty of interesting materials regarding bilinear or multilinear forms, but they are not exactly relevant in this chapter. We leave some as exercises for interested readers. In the following exercises F denotes a field, V denotes a finite dimensional F-vector space, and B denotes an F-bilinear form on $V \times V$.

10. We say that B is **symmetric** if $B(v, w) = B(w, v)$ for all $v, w \in V$.

    Let β be an (arbitrary) ordered basis for V over F. Show that B is symmetric if and only if the matrix of B with respect to β (twice) is symmetric.

11. A **quadratic form** on V is a function $q: V \to F$ such that

    $$q(v) = B(v, v)$$

    for some bilinear form B on $V \times V$.

    Assume Char $F \neq 2$. Show that for each quadratic form q on V, there is one and only one *symmetric* bilinear form B on $V \times V$ such that $q(v) = B(v, v)$ for all $v \in V$.

    When Char $F = 2$ and $\dim_F V > 1$, the correspondence between quadratic forms and symmetric bilinear forms breaks down.

12. We say that the bilinear form B is **skew-symmetric** or **antisymmetric** if $B(w, v) = -B(v, w)$ for all $v, w \in V$.

    Let β be an (arbitrary) ordered basis for V over F. Show that B is skew-symmetric if and only if the matrix of B with respect to β (twice) is skew-symmetric.

13. Assume Char $F \neq 2$. Show that any bilinear form on $V \times V$ is the sum of a symmetric bilinear form and a skew-symmetric bilinear form. (Hint: Let A be a matrix of B with respect to a certain basis twice. Consider $A^+ = (1/2)(A + A^{\mathrm{tr}})$ and $A^- = (1/2)(A - A^{\mathrm{tr}})$.)

14. We say that B is **alternating** if $B(v, v) = 0$ for all $v \in V$.

    (a) Show that any alternating bilinear form is skew-symmetric. (Hint: Expand $B(v + w, v + w)$.)

    (b) Let Char $F \neq 2$. Show that a bilinear form is alternating if and only if it is skew-symmetric.

    (c) Let Char $F = 2$. Show that B is alternating if and only if it is symmetric and $B(u, u) = 0$ for all u in some basis of V.

15. Let $\dim_F V = n$. Show that the subset of alternating bilinear forms is a subspace of $\mathrm{Bil}_F(V \times V, F)$. Determine the dimension of the subspace of alternating bilinear forms over F.

16. Let V and W be finite dimensional vector spaces over F. Show that the following conditions are equivalent for the multilinear map $f \colon V^n \to W$:

    (i) $f(v_1, v_2, \ldots, v_n) = 0$ whenever $v_i = v_{i+1}$ for some $1 \leq i \leq n-1$;

    (ii) $f(v_1, v_2, \ldots, v_n) = 0$ whenever $v_i = v_j$ for some $1 \leq i < j \leq n$;

    (iii) $f(v_1, v_2, \ldots, v_n) = 0$ whenever $v_1, v_2, \ldots, v_n$ are linearly dependent over F.

    If f satisfies one of the conditions above, we say that f is an **alternating** multilinear map over F.

5.2   Tensor products of vector spaces

The tensor product is a great and useful tool in many fields of algebra as well as in many other disciplines of mathematics. Though not difficult, it is confusing to most students who study it for the very first time. For the rest of this book we will present definitions and discuss basic properties regarding tensor products. However, we will not discuss usage and applications of tensor products for these discussions are beyond the scope of this book. We will simply stress its importance by stating that the tensor product is the *coproduct* in its relevant category.

We will start from the easiest case of a tensor product of two finite dimensional vector spaces. Throughout this section, U, V and W denote *finite dimensional* vector spaces over the field F.

The definition

Definition 5.2.1. Let U and V be finite dimensional F-vector spaces. We define the **tensor product** $U \otimes_F V$, or simply $U \otimes V$ when F is understood, to be the dual space of $\mathrm{Bil}_F(U \times V, F)$. An element in $U \otimes V$ is called a **tensor**.

Definition 5.2.2. For $u \in U$ and $v \in V$, define

$$
\begin{array}{rccc}
u \otimes v: & \mathrm{Bil}_F(U \times V, F) & \longrightarrow & F \\
& B & \longmapsto & B(u, v).
\end{array}
$$

In Lemma 5.2.3 we will show that $u \otimes v$ is an element in $U \otimes_F V$.[1] A tensor of the form $u \otimes v$ is also called an **elementary tensor** or a **decomposable tensor**.

Lemma 5.2.3. *For $u \in U$ and $v \in V$, the map $u \otimes v$ is an F-linear map. Hence $u \otimes v \in U \otimes_F V$.*

Proof. Let $a_1, a_2 \in F$ and $B_1, B_2 \in \mathrm{Bil}_F(U \times V, F)$. We have that

$$
(u \otimes v)(a_1 B_1 + a_2 B_2) = (a_1 B_1 + a_2 B_2)(u, v)
$$
$$
= a_1 B_1(u, v) + a_2 B_2(u, v) = a_1(u \otimes v)(B_1) + a_2(u \otimes v)(B_2).
$$

Hence $u \otimes v$ is F-linear. $\qquad\square$

As a dual space, $U \otimes V$ is a vector space over F. The following lemma gives the basic properties regarding the *tensors*.

Lemma 5.2.4. *The following identities hold in $U \otimes_F V$:*

(i) $u \otimes (v + v') = u \otimes v + u \otimes v'$,

(ii) $(u + u') \otimes v = u \otimes v + u' \otimes v$, *and*

(iii) $(au) \otimes v = a(u \otimes v) = u \otimes (av)$

---

[1] I read $u \otimes v$ as "u tensor v", in case anyone is interested.

for all $a \in F$, u, $u' \in U$ *and* v, $v' \in V$.

Proof. All the terms in this lemma are maps. It suffices to compare values on both sides. Let $a \in F$, u, $u' \in U$ and v, $v' \in V$. We have

$$\begin{aligned}
\text{(i)} \quad & \big(u \otimes (v + v')\big)(B) = B(u, v + v') = B(u, v) + B(u, v') \\
& = (u \otimes v)(B) + (u \otimes v')(B) = (u \otimes v + u \otimes v')(B), \\
\text{(ii)} \quad & \big((u + u') \otimes v\big)(B) = B(u + u', v) = B(u, v) + B(u', v) \\
& = (u \otimes v)(B) + (u' \otimes v)(B) = (u \otimes v + u' \otimes v)(B), \\
\text{(iii)} \quad & \big((au) \otimes v\big)(B) = B(au, v) = aB(u, v) = a(u \otimes v)(B) \\
& = \big(a(u \otimes v)\big)(B), \text{ and} \\
& \big(u \otimes (av)\big)(B) = B(u, av) = aB(u, v) = a(u \otimes v)(B) \\
& = \big(a(u \otimes v)\big)(B)
\end{aligned}$$

for any $B \in \mathrm{Bil}_F(U \times V, F)$. Hence the result. $\square$

Sadly, not every tensor in $U \otimes V$ is decomposable (see Exercise 2). However, for finite dimensional vector spaces U and V, we may find a basis consisting of decomposable tensors for $U \otimes V$ as we shall see in the following theorem.

Theorem 5.2.5. *Suppose U and V are finite dimensional vector spaces and suppose $\beta = (u_1, u_2, \ldots, u_m)$ and $\gamma = (v_1, v_2, \ldots, v_n)$ are ordered F-bases for U and V respectively. Then*

$$\mathscr{B} = \big\{ u_i \otimes v_j \in U \otimes V : i = 1, \ldots, m, \ j = 1, \ldots, n \big\}$$

is a basis for $U \otimes V$ over F. Hence $\dim_F U \otimes V = mn$.

Remark. In particular, we can see that $U \otimes V$ is spanned by the decomposable tensors. It turns out that for many of the results in this section, it suffices to examine the decomposable tensors only.

Proof. Let $\{B_{ij}\}_{\substack{1 \leq i \leq m \\ 1 \leq j \leq n}}$ be as given in (5.1.2). From Proposition 5.1.10 it is an F-basis for $\mathrm{Bil}_F(U \times V, F)$. From Exercise 15, §2.1, the set $\{f_{ij}\}_{\substack{1 \leq i \leq m \\ 1 \leq j \leq n}}$ in the dual space of $\mathrm{Bil}_F(U \times V)$ where

$$f_{ij}(B_{k\ell}) = \begin{cases} 1, & \text{if } k = i \text{ and } \ell = j, \\ 0, & \text{otherwise,} \end{cases}$$

is an F-basis for $U \otimes V$. Note that

$$u_i \otimes v_j(B_{k\ell}) = B_{k\ell}(u_i,\, v_j) = \begin{cases} 1, & \text{if } k = i \text{ and } \ell = j, \\ 0, & \text{otherwise,} \end{cases}$$

by (5.1.2). Since $u_i \otimes v_j$ agrees with f_{ij} on an F-basis of $\mathrm{Bil}_F(U \times V,\, F)$, we have that $u_i \otimes v_j = f_{ij}$ for all i and j. In other words, $\mathscr{B}$ is indeed an F-basis for $U \otimes V$. $\qquad\square$

The universal property of the tensor product

Next we will give the *universal property* of the tensor product, the property that describes the tensor product definitely.

Proposition 5.2.6. *There is a canonical F-bilinear map $\mathfrak{b} \colon U \times V \to U \otimes V$ sending $(u,\, v)$ to $u \otimes v$.*

Proof. From Lemma 5.2.4 we have

$$\mathfrak{b}(au + a'u',\, v) = (au + a'u') \otimes v = a(u \otimes v) + a'(u' \otimes v)$$
$$= a\mathfrak{b}(u,\, v) + a'\mathfrak{b}(u',\, v),$$
$$\mathfrak{b}(u,\, av + a'v') = u \otimes (av + a'v') = a(u \otimes v) + a'(u \otimes v')$$
$$= a\mathfrak{b}(u,\, v) + a'\mathfrak{b}(u,\, v')$$

for $a,\, a' \in F$, $u,\, u' \in U$ and $v,\, v' \in V$. Hence $\mathfrak{b}$ is bilinear over F. $\qquad\square$

Theorem 5.2.7 (The universal property of the tensor product). *Let U, V and W be vector spaces over F where U and V are finite dimensional. Suppose B is an F-bilinear map from $U \times V$ to W. There exists a unique F-linear map L form $U \otimes V$ to W such that $B = L\mathfrak{b}$.*

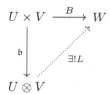

Proof. Let $\{u_1, u_2, \ldots, u_m\}$ and $\{v_1, v_2, \ldots, v_n\}$ be F-bases for U and V respectively. By Theorem 5.2.5, $\mathscr{B} = \{u_i \otimes v_i\}_{i,j}$ is a basis for $U \otimes V$ over F. If the linear map L exists, it is required that

$$L(u_i \otimes v_j) = L\flat(u_i, v_j) = B(u_i, v_j)$$

for all i and j. Hence L is unique if it exists.

From Theorem 2.1.11, there is a linear map $L: U \otimes V \to W$ such that $L(u_i \otimes v_j) = B(u_i, v_j)$ for all i and j. It remains to show that $B = L\flat$.

Let $u = \sum_i a_i u_i$ and $v = \sum_j b_j v_j$ where $a_i, b_j \in F$. Then

$$
\begin{aligned}
u \otimes v &= \left(\sum_i a_i u_i\right) \otimes \left(\sum_j b_j v_j\right) \\
&= \sum_i a_i \left(u_i \otimes \left(\sum_j b_j v_j\right)\right), \qquad \text{by Lemma 5.2.4(ii),} \\
&= \sum_i a_i \left(\sum_j b_j (u_i \otimes v_j)\right), \qquad \text{by Lemma 5.2.4(i),} \\
&= \sum_{i,j} a_i b_j (u_i \otimes v_j).
\end{aligned}
$$

(5.2.1)

It follows that

$$
\begin{aligned}
L\flat(u, v) = L(u \otimes v) &= L\left(\sum_{i,j} a_i b_j u_i \otimes v_j\right), \qquad \text{by (5.2.1),} \\
&= \sum_{i,j} a_i b_j L(u_i \otimes v_j) \\
&= \sum_{i,j} a_i b_j B(u_i, v_j), \qquad \text{by the definition of } L, \\
&= B\left(\sum_i a_i u_i, \sum_j b_j v_j\right), \qquad \text{since } B \text{ is bilinear,} \\
&= B(u, v).
\end{aligned}
$$

Thus $B = L\flat$. $\qquad\qquad\qquad\qquad\qquad\qquad\qquad\qquad\qquad\qquad\qquad$ $\square$

We may use the universal property to verify some of the fundamental properties regarding tensor products.

Proposition 5.2.8. *The following properties hold for all finite dimensional F-vector spaces U, V and W:*

(i) (**Commutativity**) $U \otimes V \cong V \otimes U$;

(ii) (**Associativity**) $(U \otimes V) \otimes W \cong U \otimes (V \otimes W)$;

(iii) (**Distributivity**) $(U \oplus V) \otimes W \cong (U \otimes W) \oplus (V \otimes W)$.

Proof. (i) Let $B \colon U \times V \to V \otimes U$ be the map sending (u, v) to $v \otimes u$. It is routine to check that B is F-bilinear. By Theorem 5.2.7, there is a linear map from $U \otimes V$ to $V \otimes U$ sending $u \otimes v$ to $v \otimes u$. By symmetry, there is a linear map L' from $V \otimes U$ to $U \otimes V$ sending $v \otimes u$ to $u \otimes v$. Note that $L'L$ sends $u \otimes v$ to $u \otimes v$ for all $u \in U$ and $v \in V$, which agrees with $1_{U \otimes V}$ on a set of generators for $U \otimes V$. Hence $L'L = 1_{U \otimes V}$. By symmetry, $LL' = 1_{V \otimes U}$. Hence L is an isomorphism.

(ii) For each u in U, we will define a map

$$
\begin{aligned}
B_u \colon \quad V \times W &\longrightarrow (U \otimes V) \otimes W \\
(v, w) &\longmapsto (u \otimes v) \otimes w
\end{aligned}
\quad .
$$

It is routine to check that B_u is bilinear for any $u \in U$. By Theorem 5.2.7, there exists a linear map $L_u \colon V \otimes W \to (U \otimes V) \otimes W$ such that $B_u = L_u \flat_{V, W}$ where $\flat_{V, W}$ is the canonical map from $V \times W$ to $V \otimes W$. In particular,

$$
L_u(v \otimes w) = L_u \flat_{V, W}(v, w) = B_u(v, w) = (u \otimes v) \otimes w.
$$

Now define another map

$$
\begin{aligned}
B \colon \quad U \times (V \otimes W) &\longrightarrow (U \otimes V) \otimes W \\
(u, x) &\longmapsto L_u(x)
\end{aligned}
\quad .
$$

To check that B is bilinear, first note that $B(u, -) = L_u$ is linear. It remains to check that $B(-, x)$ is linear for any $x \in V \otimes W$. For this we need to check the identity

$$
(5.2.2) \qquad\qquad L_{bu + bu'} = bL_u + b'L_{u'}
$$

holds for b, $b' \in F$ and u, $u' \in U$. To check the two linear maps are equal, we only need to compare values of both sides on the decomposable tensors. We have

$$
L_{bu + b'u}(v \otimes w) = \big((bu + b'u') \otimes v\big) \otimes w
$$

$$= \big((bu) \otimes v\big) \otimes w + \big((b'u') \otimes v\big) \otimes w$$
$$= \big(b(u \otimes v)\big) \otimes w + \big(b'(u' \otimes v)\big) \otimes w$$
$$= b\big((u \otimes v) \otimes w\big) + b'\big((u' \otimes v) \otimes w\big)$$
$$= bL_u(v \otimes w) + b'L_{u'}(v \otimes w)$$

for all v and $w \in W$. Hence the identity in (5.2.2) holds. We now have

$$B\left(bu + b'u_i,\, x\right) = L_{bu+b'u'}\left(x\right) = bL_u\left(x\right) + bL_{u'}\left(x\right)$$
$$= bB\left(u,\, x\right) + b'B\left(u',\, x\right)$$

for $b,\, b' \in F$, $u,\, u' \in U$ and $x \in V \otimes W$. We have shown that B is bilinear. From Theorem 5.2.7, we have an F-linear map

$$
\begin{array}{rccc}
L: & U \otimes (V \otimes W) & \longrightarrow & (U \otimes V) \otimes W \\
& u \otimes (v \otimes w) & \longmapsto & (u \otimes v) \otimes w
\end{array}.
$$

By symmetry we also have an F-linear map

$$
\begin{array}{rccc}
L': & (U \otimes V) \otimes W & \longrightarrow & U \otimes (V \otimes W) \\
& (u \otimes v) \otimes w & \longmapsto & u \otimes (v \otimes w)
\end{array}.
$$

Clearly the two maps L and L' are inverse to each other. Hence L is an isomorphism.

(iii) We leave this as an exercise.   □

Remark. Since $\otimes$ satisfies associativity, we may use $U \otimes V \otimes W$ to denote either $(U \otimes V) \otimes W$ or $U \otimes (V \otimes W)$.

The universal property of the tensor product practically "defines" it, as we shall see in the following result.

Theorem 5.2.9. *Suppose X is an F-vector space satisfying the universal property described in Theorem 5.2.7. To be more precise, suppose there is a canonical bilinear map $\flat^\sharp$ from $U \times V$ to X such that for any bilinear map B from $U \times V$ to W, there exists a unique linear map L from X to W such that $B = L\flat^\sharp$. Then there is a canonical isomorphism ℓ from $U \otimes V$ to X satisfying $\flat^\sharp = \ell\flat$ where $\flat$ is the canonical bilinear map from $U \times V$ to $U \otimes V$.*

Proof. From Theorem 5.2.7, there is a unique linear map ℓ from $U \otimes V$ to X such that $\flat^\sharp = \ell\flat$. From assumption, there is also a unique linear map ℓ' from X to $U \otimes V$ such that $\flat = \ell'\flat^\sharp$.

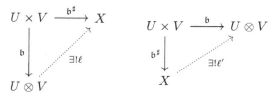

Note that

$$\flat^\sharp = \ell\flat = (\ell\ell')\flat^\sharp \quad \text{and} \quad \flat = \ell'\flat^\sharp = (\ell'\ell)\flat.$$

However, from assumption the linear map $\mathbf{1}_X$ should be the unique linear map such that $\flat^\sharp = \mathbf{1}_X \flat^\sharp$. Form Theorem 5.2.7, $\mathbf{1}_{U \otimes V}$ should be the unique linear map such that $\flat = \mathbf{1}_{U \otimes V}\flat$.

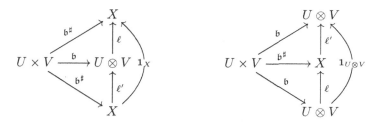

Hence $\ell\ell' = \mathbf{1}_X$ and $\ell'\ell = \mathbf{1}_{U \otimes V}$. We conclude that ℓ is an isomorphism from $U \otimes V$ to X and it is the unique isomorphism from $U \otimes V$ to X such that $\flat^\sharp = \ell\flat$. □

By virtue of the theorem above, we may make the following definition.

Definition 5.2.10. If $\flat^\sharp$ is an R-bilinear map from $U \times V$ to X satisfying the universal property of the tensor product, we also call the pair $(X, \flat^\sharp)$ the **tensor product** of U and V (up to canonical isomorphism).

Unfortunately, the treatment of tensor products in this section only applies to the simplest case (of finite dimensional vector spaces) and it often leads to unpleasant technical complications. Worst of all, it cannot be readily generalized. In the next section, we will develop a different approach to tensor products.

Exercises 5.2

In the following exercises, all the tensor products are over the field F.

1. Let V be a finite dimensional F-vector space and let u, $v \in V$. Determine whether $u \otimes v = v \otimes u$ in $V \otimes V$ in general?

2. Show that $e_1 \otimes e_2 + e_2 \otimes e_1$ is not decomposable in $F^2 \otimes F^2$.

3. Prove Proposition 5.2.8(iii), the *distributivity* of tensor products.

4. Let m and n be non-negative integers and let x and y be indeterminates over F. Remember that we may use $\mathscr{P}_n$ to denote the subspace generated by monomials of degree $\leq n$ in $F[x]$. We now will use $\mathscr{P}_{m,n}$ to denote the subspace of $F[x,y]$ generated by monomials whose degree in x is $\leq m$ and whose degree in y is $\leq n$. For example, $\mathscr{P}_{1,2}$ is generated by $\{1, y, y^2, x, xy, xy^2\}$ over F.

   Show that the map $\flat^\sharp \colon \mathscr{P}_m \times \mathscr{P}_n \to \mathscr{P}_{m,n}$ sending $(f(x), g(x))$ to $f(x)g(y)$ is a bilinear map over F. Show that $(\mathscr{P}_{m,n}, \flat^\sharp)$ is the tensor product $\mathscr{P}_m \otimes_F \mathscr{P}_n$.

5. Let U, V, W and X be F-vector spaces. Define $U \otimes'_F V \otimes'_F W$ (F may be omitted when it is understood) to be the dual space of $\mathrm{Tril}_F(U \times V \times W, X)$. Is it true that $U \otimes' V \otimes' W \cong U \otimes V \otimes W$?

5.3 Tensor products of modules

There is a second approach to tensor products which may seem abstract compared with the concrete construction in §5.2. However this different approach applies in greater generality and is no more difficult[2] than the first approach.

In this section we will define tensor products of modules. Since a vector space is just a special type of module, this will indeed be a generalization. The good news is that the approach in this section will apply to infinite dimensional vector spaces as well. We will first define tensor products by

---

[2]This is my wish.

using "formal" products. Then we will establish the universal property of tensor products by following a similar route as in §5.2. By Theorem 5.2.9, this "new" type of tensor product of two finite dimensional vector spaces will be isomorphic to the tensor product defined in the previous section.

Throughout this section R denotes a ring, and M, N denote arbitrary R-modules.

The construction

Let $\mathfrak{F}$ be the free module on the basis

$$\{e(m, n) : m \in M, \ n \in N\}$$

over R. This might be slightly difficult to comprehend for a first-timer. We will try to use an example to demonstrate what $\mathfrak{F}$ is. Suppose $R = \mathbb{Z}$ and $M = N = \mathbb{Z}_2$. Then

$$M \times N = \{(0, 0), \ (0, 1), \ (1, 0), \ (1, 1)\}$$

is a set of four elements. In this case, $\mathfrak{F}$ is a free $\mathbb{Z}$-module of rank 4. The elements in $\mathfrak{F}$ are simply "formal" linear combinations of the form

$$ae(0, 0) + be(0, 1) + ce(1, 0) + de(1, 1), \qquad a, b, c, d \in \mathbb{Z}.$$

However, more often than not, M and N are infinite sets. So $\mathfrak{F}$ is usually a very "large" free module.

Let K be the R-submodule of $\mathfrak{F}$ generated by *all* elements of the form

(5.3.1)
$$e(am + a'm', n) - ae(m, n) - a'e(m', n), \quad \text{or}$$
$$e(m, an + a'n') - ae(m, n) - a'e(m, n')$$

for a, a' in R, m, m' in M and n, n' in N. We will define $M \otimes_R N$, or simply $M \otimes N$ when R is understood, to be the quotient module $\mathfrak{F}/K$. We will use $m \otimes n$ to denote the coset of $e(m, n)$. Using this notation, (5.3.1) may be translated as the identities

(5.3.2)
$$(am + a'm') \otimes n = a(m \otimes n) + a'(m' \otimes n), \quad \text{and}$$
$$m \otimes (an + a'n') = a(m \otimes n) + a'(m \otimes n')$$

for a, a' in R, m, m' in M and n, n' in N.

The universal property

Proposition 5.3.1. *Let $\flat$ be the mapping from $M \times N$ to $M \otimes N$ sending (m, n) to $m \otimes n$. The map $\flat$ is bilinear over R.*

Proof. The result follows from the identities in (5.3.2). □

We will call the map $\flat$ in Proposition 5.3.1 the **canonical map** from $M \times N$ to $M \otimes N$.

A tensor of the form $m \otimes n$ may still be called an **elementary tensor** or a **decomposable tensor**. It is a fact that $M \otimes N$ is generated by all the decomposable tensors since in the construction of $M \otimes N$, $\mathfrak{F}$ is generated by elements of the form $e(m, n)$ for $m \in M$ and $n \in N$.

With the new approach, it is just as easy to verify the universal property of the more general type of tensor product.

Theorem 5.3.2 (The universal property of the tensor product). *Let W be an R-module and let $\flat$ be the canonical map from $M \times N$ to $M \otimes_R N$. Suppose B is an R-bilinear map from $M \times N$ to W. There exists a unique R-linear map L form $M \otimes N$ to W such that $B = L\flat$.*

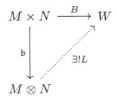

Moreover, there is a bijection between

$$\mathrm{Bil}_R(M \times N, W) \qquad and \qquad \mathrm{Hom}_R(M \otimes N, W).$$

Proof. Since $L(m \otimes n) = L\flat(m, n) = B(m, n)$ for all $(m, n) \in M \times N$ if L exists, the value of L is determined on a generating set of $M \otimes N$. Hence L is unique if it exists. Our goal is to show that L exists.

By Theorem 2.1.11, there exists an R-linear map f from $\mathfrak{F}$ to W sending $e(m, n)$ to $B(m, n)$ for all $m \in M$ and $n \in N$. Since B is R-bilinear,

$$f\big(e(am + a'm', n) - ae(m, n) - a'e(m', n)\big)$$
$$= f\big(e(am + a'm', n)\big) - af\big(e(m, n)\big) - a'f\big(e(m', n)\big)$$

$$= B(am + a'm', n) - aB(m, n) - a'B(m', n) = 0;$$
$$f\big(e(m, an + a'n') - ae(m, n) - a'e(m, n')\big)$$
$$= f\big(e(m, an + a'n')\big) - af\big(e(m, n)\big) - a'f\big(e(m, n')\big)$$
$$= B(m, an + a'n') - aB(m, n) - a'B(m, n') = 0$$

for a, a' in R, m, m' in M and n, n' in N. This shows that $K \subseteq \operatorname{Ker} f$. By the fundamental theorem of linear maps (Theorem 2.2.6), there is a unique R-linear map L from $M \otimes N = \mathfrak{F}/K$ to W sending $m \otimes n$ to $B(m, n)$. This says that $B = L\flat$.

In the previous paragraph we have constructed a mapping

$$\phi: \quad \operatorname{Bil}_R(M \times N, W) \quad \longrightarrow \quad \operatorname{Hom}_R(M \otimes N, W)$$
$$B \quad \longmapsto \quad \phi(B): m \otimes n \mapsto B(m, n) \quad.$$

To show that ϕ is a one-to-one correspondence, it suffices to show that the mapping

$$\psi: \quad \operatorname{Hom}_R(M \otimes N, W) \quad \longrightarrow \quad \operatorname{Bil}_R(M \times N, W)$$
$$L \quad \longmapsto \quad L\flat$$

is the inverse of ϕ. Note that $L\flat \in \operatorname{Bil}_R(M \times N, W)$ from Exercise 1, §5.1.

Let $B \in \operatorname{Bil}_R(M \times N, W)$. For any $(m, n) \in M \times N$ we have

$$\psi\big(\phi(B)(m, n)\big) = \phi(B)\big(\flat(m, n)\big) = \phi(B)(m \otimes n) = B(m, n).$$

This shows that $\psi\varphi(B) = B$. Conversely, let $L \in \operatorname{Hom}_R(M \otimes N, W)$. For any $m \in M$ and $n \in N$, we have

$$\phi\big(\psi(L)(m \otimes n)\big) = \psi(L)(m, n) = L\flat(m, n) = L(m \otimes n).$$

Since $\phi\psi(L)$ and L agree on a set of generators of $M \otimes N$, we conclude that $\phi\psi(L) = L$. Hence, ϕ and ψ are inverse to each other. $\qquad\square$

In §5.2, several results on the tensor product was proven using the universal property only. The proofs are valid verbatim for their module counterparts. We will give these results below.

Theorem 5.3.3. *Let M and N be R-modules. Suppose X is an R-module satisfying the universal property of the tensor product. To be precise, suppose there is a canonical bilinear map $\flat^\sharp$ from $M \times N$ to X such that for* *any bilinear map B from M × N to W, there exists a unique linear map L from X to W such that B = Lb♯. Then there is a canonical isomorphism ℓ from M ⊗ N to X satisfying b♯ = ℓb where b is the canonical map from M × N to M ⊗ N.*

Cf. Theorem 5.2.9.

Definition 5.3.4. Let M, N and X be R-modules. If $b^\sharp$ is an R-bilinear map from $M \times N$ to X satisfying the universal property of the tensor product of modules, we also call the pair $(X, b^\sharp)$ the **tensor product** of M and N (up to canonical isomorphism) over R.

Here is another result needing no proof.

Proposition 5.3.5. *The following properties hold for all R-modules M, N and W.*

(i) (**Commutativity**) $M \otimes N \cong N \otimes M$;

(ii) (**Associativity**) $(M \otimes N) \otimes W \cong M \otimes (N \otimes W)$;

(iii) (**Distributivity**) $(M \oplus N) \otimes W \cong (M \otimes W) \oplus (N \otimes W)$.

Cf. Proposition 5.2.8. Note that we no longer require the modules be free or finitely generated.

With the construction in this section, it is now possible for us to generalize the *distributivity* of $\otimes$ over $\oplus$ in Proposition 5.3.5(iii).

Proposition 5.3.6. *Let I be an arbitrary index set. Let W and M_i, $i \in I$, be R-modules. Then $\left(\bigoplus_i M_i \right) \otimes W \cong \bigoplus_i (M_i \otimes W)$.*

Proof. Define the map $B \colon \left(\bigoplus_i M_i \right) \times W \to \bigoplus_i (M_i \otimes W)$ by letting

$$B\big((m_i)_i,\, w\big) = (m_i \otimes w)_i.$$

It is routine to verify that B is bilinear over R. By Theorem 5.3.2, there is an R-linear map L from $\left(\bigoplus_i M_i \right) \otimes W$ to $\bigoplus_i (M_i \otimes W)$ such that

$$L\big((m_i)_i \otimes w\big) = (m_i \otimes w)_i.$$

Before we continue, let's set up some notations to facilitate our discussion. There is a natural injective R-linear map η_{i_0} from M_{i_0} to $\bigoplus_i M_i$ for each i_0 by sending x to $(m_i)_i$ where $m_{i_0} = x$ and $m_i = 0$ for all $i \neq i_0$. In other words, $\eta_{i_0}(x)$ has at most one nonzero coordinate and any nonzero coordinate must appear at the i_0-th spot. Suppose given $(m_i)_{i \in I} \in \bigoplus_{i \in I} M_i$ such that $m_i = 0$ except for $i_1, i_2, \ldots, i_n$. Then

$$(m_i)_{i \in I} = \sum_{j=1}^{n} \eta_{i_j}(m_{i_j}).$$

Similarly, we will use $\widetilde{\eta}_{i_0}$ to send an element of $M_{i_0} \otimes W$ to an element in $\bigoplus_i (M_i \otimes W)$ whose i_0-th coordinate is the given element and whose other coordinates are zero. The map $\widetilde{\eta}_{i_0}$ is also easily seen to be R-linear.

Define the map $b_{i_0} \colon M_{i_0} \times W \to \left(\bigoplus_i M_i \right) \otimes W$ by letting

$$b_{i_0}(x, w) = \eta_{i_0}(x) \otimes w \qquad \text{for } x \in M_{i_0} \text{ and } w \in W.$$

It is again routine to verify that b_{i_0} is bilinear over R for all i_0. By Theorem 5.3.2, there is an R-linear map ℓ_{i_0} from $M_{i_0} \otimes W$ to $\left(\bigoplus_i M_i \right) \otimes W$ sending $x \otimes w$ to $\eta_{i_0}(x) \otimes w$ for all i_0. By Exercise 14, §2.2, we now have a R-linear map L' from $\bigoplus_i (M_i \otimes W)$ to $\left(\bigoplus_i M_i \right) \otimes W$ such that

$$L'\big((\tau_i)_{i \in I}\big) = L'\left(\sum_{j=1}^{n} \widetilde{\eta}_{i_j}(\tau_{i_j}) \right) = \sum_{j=1}^{n} L'\big(\widetilde{\eta}_{i_j}(\tau_{i_j})\big) = \sum_{j=1}^{n} \ell_{i_j}(\tau_{i_j})$$

where $(\tau_i)_{i \in I} \in \bigoplus_i (M_i \otimes W)$ be such that $\tau_i = 0$ except for $i = i_1, i_2, \ldots, i_n$. The various maps in this proof are depicted in the following diagram.

Next we will check that L and L' are inverse to each other. Remember that $\left(\bigoplus_i M_i \right) \otimes W$ is generated by elements of the form $(m_i)_i \otimes w$. Let $m_i = 0$ except for $i = i_1, i_2, \ldots, i_n$. Then

$$L'L\big((m_i)_i \otimes w\big) = L'\big((m_i \otimes w)_i\big) = \sum_{j=1}^{n} \ell_{i_j}(m_{i_j} \otimes w)$$

$$= \sum_{j=1}^{n} \big(\eta_{i_j}(m_{i_j}) \otimes w\big) = \left(\sum_{j=1}^{n} \eta_{i_j}(m_{i_j}) \right) \otimes w = (m_i)_i \otimes w.$$

The R-linear map $L'L$ and $\mathbf{1}_{\left(\bigoplus_i M_i\right)\otimes W}$ agree on a generating set. Hence $L'L = \mathbf{1}_{\left(\bigoplus_i M_i\right)\otimes W}$.

Conversely, the module $\bigoplus_i(M_i \otimes W)$ is generated by elements of at most one nonzero coordinate. That is, it is generated by elements of the form $\widetilde{\eta}_{i_0}(\tau)$ for some i_0 and $\tau \in M_{i_0} \otimes W$. Since $M_{i_0} \otimes W$ is generated by $x \otimes w$ for $x \in M_{i_0}$ and $w \in W$ and $\widetilde{\eta}_{i_0}$ is R-linear, we further deduce that $\bigoplus_i(M_i \otimes W)$ is generated by

$$\left\{\widetilde{\eta}_{i_0}(x \otimes w) \in \bigoplus_i(M_i \otimes W) : i_0 \in I,\ x \in M_{i_0},\ w \in W\right\}.$$

We have that

$$LL'\big(\widetilde{\eta}_{i_0}(x \otimes w)\big) = L\big(\ell_{i_0}(x \otimes w)\big) = L\big(\eta_{i_0}(x) \otimes w\big) = \widetilde{\eta}_{i_0}(x \otimes w).$$

Hence $LL' = \mathbf{1}_{\bigoplus_i(M_i\otimes W)}$ since their values agree on a generating set. We conclude that L is an isomorphism. $\qquad\square$

How to evaluate tensor products

In this subsection we describe how to evaluate a tensor product.

Proposition 5.3.7. *Let M be an R-module. The map $R \otimes_R M \longrightarrow M$ sending $r \otimes m$ to m is an isomorphism.*

Proof. The map $R \times M \longrightarrow M$ sending (r, m) to rm is clearly bilinear over R. Thus there is an R-linear map $f\colon R \otimes M \to M$ sending $r \otimes m$ to rm. Conversely, there is clearly an R-linear map $g\colon M \to R \otimes M$ sending m to $1 \otimes m$. For $r \in R$ and $m \in M$, we have

$$gf(r \otimes m) = g(rm) = 1 \otimes rm = r \otimes m.$$

Thus $gf = \mathbf{1}_{R\otimes M}$ since their values agree on all the decomposable tensors. Conversely,

$$fg(m) = f(1 \otimes m) = 1m = m.$$

Thus $fg = \mathbf{1}_M$. It follows that f is an isomorphism. $\qquad\square$

Corollary 5.3.8. *Let I and J be index sets. Let M, $\{M_i\}_{i\in I}$ and $\{N_j\}_{j\in J}$ be R-modules.*

(a) We have $\left(\bigoplus_{i \in I} M_i \right) \otimes \left(\bigoplus_{j \in J} N_j \right) \cong \bigoplus_{i \in I, j \in J} (M_i \otimes N_j)$.

(b) Let $F = \bigoplus_{i \in I} Re_i$ be free on the basis $\{e_i\}_{i \in I}$ over R. Then

$$F \otimes M \cong \bigoplus_{i \in I} (Re_i \otimes M) \cong \bigoplus_{i \in I} M.$$

(c) Let $G = \bigoplus_{j \in J} Re'_j$ be free on the basis $\{e'_j\}_{j \in J}$ over R. Then $F \otimes G$ is
free on the basis $\{e_i \otimes e'_j : i \in I, j \in J\}$ over R.

Proof. (a) This follows from Proposition 5.3.6 and Proposition 5.3.5(i).

(b) This follows from Proposition 5.3.6 and Proposition 5.3.7.

(c) This follows from (a) and the fact that $Re_i \otimes Re'_j = Re_i \otimes e'_j$ is free
by Proposition 5.3.7.   □

Corollary 5.3.8(c) generalizes Theorem 5.2.5 considerably, from finite
dimensional vector spaces to free modules of arbitrary rank.

Example 5.3.9. In this example we will roughly explain why the set of
decomposable tensors is relatively small in a tensor product.

Let M be a free R-module on the basis $\{e_i\}_{i=1}^m$ and N be a free R-module
on the basis $\{f_j\}_{j=1}^n$. Then $M \otimes N$ is a free module of rank mn on the basis
$\{e_i \otimes f_j\}_{i,j}$. A typical element in $M \otimes N$ is of the form $\sum_{i,j} a_{ij} e_i \otimes f_j$, which
corresponds uniquely to a matrix (a_{ij}) in $M_n(R)$. (The correspondence is
not canonical: it depends on the basis chosen.) A decomposable tensor is
of the form

$$\left(\sum_i a_i e_i \right) \otimes \left(\sum_j b_j f_j \right) = \sum_{i,j} a_i b_j e_i \otimes f_j,$$

which corresponds to the matrix

$$\begin{pmatrix} a_1 \\ a_2 \\ \vdots \\ a_m \end{pmatrix} \begin{pmatrix} b_1 & b_2 & \cdots & b_n \end{pmatrix}.$$

From Exercise 1, §3.5, the rank of such a matrix is at most one, which
makes it quite special among the square matrices of size n.

To better understand tensor products, the following result is useful.

Proposition 5.3.10. *Let M, M', N and N' be R-modules. Let $f\colon M \to N$ and $g\colon M' \to N'$ be two R-linear maps. There exists an R-linear map $f \otimes g\colon M \otimes N \to M' \otimes N'$ such that*

$$(f \otimes g)(m \otimes m') = f(m) \otimes g(m')$$

for all $m \in M$ and $m' \in M'$.

Proof. Let $\flat$ be the canonical map from $M \times M'$ to $M \otimes M'$. Consider the map $f \times g\colon M \times M' \to N \otimes N'$ sending (m, m') to $f(m) \otimes g(m')$. It is routine to check that $f \times g$ is bilinear. Thus, there exists an R-linear map $f \otimes g$ from $M \otimes M'$ to $N \otimes N'$ such that $f \times g = (f \otimes g) \circ \flat$. We have that

$$(f \otimes g)(m \otimes m') = (f \otimes g)\big(\flat(m, m')\big) = (f \times g)(m, n) = f(m) \otimes g(m')$$

for all $m \in M$ and $m' \in M'$. $\square$

Proposition 5.3.11. *Let N be an R-submodule of M and let W be an R-module. We will use $\iota\colon N \to M$ to denote the inclusion map from N into M. From Proposition 5.3.10, there is an R-linear map $\iota \otimes 1_W$ from $N \otimes W$ to $M \otimes W$. Then $(M/N) \otimes W \cong (M \otimes W)/\operatorname{Im}(\iota \otimes 1_W)$.*

Proof. To simplify notation, we will use K to denote $\operatorname{Im}(\iota \otimes 1_W)$, the image of $N \otimes W$ inside $M \otimes W$.

First we verify that there is a well-defined map B from $(M/N) \times W$ to $(M \otimes W)/K$ by sending $(m + N, w)$ to $(m \otimes w) + K$. Let $m + N = m' + N$. We may find $n \in N$ such that $m' = m + n$. Then

$$m' \otimes w = (m + n) \otimes w = m \otimes w + n \otimes w,$$

where $n \otimes w \in K$. It follows that $(m \otimes w) + K = (m' \otimes w) + K$. Hence the map B is well-defined. It is routine to check that B is also bilinear over R. We may thus obtain an R-linear map $f\colon (M/N) \otimes W \to (M \otimes W)/K$ sending $(m + N) \otimes w$ to $(m \otimes w) + K$.

Conversely, let π be the canonical map form M onto M/N. From Proposition 5.3.10, there is an R-linear map $\pi \otimes 1_W$ sending $m \otimes w$ to $(m + N) \otimes w$. The submodule K is generated by $n \otimes w$ in $M \otimes W$. Note that $(\pi \otimes 1_W)(n \otimes w) = (n + N) \otimes w = 0$ in $(M/N) \otimes W$. This shows that

$K \subseteq \mathrm{Ker}(\pi \otimes 1_W)$. From Theorem 2.2.6, there is an R-linear map g from $(M \otimes N)/K$ to $(M/N) \otimes W$ sending $m \otimes w + K$ to $(m + N) \otimes w$.

It is easy to see that f and g are inverse to each other. Hence f is an isomorphism. □

Let M and W be R-modules and let $N \xrightarrow{\iota} M$ be the inclusion map. We will also use $\mathrm{Im}(N \otimes W)$ to denote the image of $\iota \otimes 1_W$. Next we will use an example to demonstrate that $n \otimes w$, $n \in N$ and $w \in W$, may have different meanings in $N \otimes W$ and in $\mathrm{Im}(N \otimes W)$.

Example 5.3.12. Consider $N = 2\mathbb{Z}$ and $W = \mathbb{Z}/2\mathbb{Z}$ as $\mathbb{Z}$-modules. From Proposition 5.3.7 we have

$$2\mathbb{Z} \otimes_{\mathbb{Z}} \frac{\mathbb{Z}}{2\mathbb{Z}} \cong \mathbb{Z} \otimes_{\mathbb{Z}} \frac{\mathbb{Z}}{2\mathbb{Z}} \cong \frac{\mathbb{Z}}{2\mathbb{Z}} \neq 0$$

since $2\mathbb{Z}$ is free on the basis $\{\bar{2}\}$. Note that $2\mathbb{Z} \otimes_{\mathbb{Z}} (\mathbb{Z}/2\mathbb{Z})$ is generated by $2 \otimes_{\mathbb{Z}} \bar{1}$. Hence $2 \otimes_{\mathbb{Z}} \bar{1} \neq 0$ in $2\mathbb{Z} \otimes_{\mathbb{Z}} (\mathbb{Z}/2\mathbb{Z})$. On the other hand,

$$2 \otimes \bar{1} = 1 \otimes 2 \cdot \bar{1} = 1 \otimes \bar{0} = 0$$

in $\mathbb{Z} \otimes_{\mathbb{Z}} (\mathbb{Z}/2\mathbb{Z})$. In fact, $\mathrm{Im}\big(2\mathbb{Z} \otimes_{\mathbb{Z}} (\mathbb{Z}/2\mathbb{Z})\big)$ is trivial in $\mathbb{Z} \otimes_{\mathbb{Z}} (\mathbb{Z}/2\mathbb{Z})$. What makes the difference is that $1 \otimes 2 \cdot \bar{1}$ is not an element in $2\mathbb{Z} \otimes_{\mathbb{Z}} (\mathbb{Z}/2\mathbb{Z})$. Observe that

$$2\mathbb{Z} \otimes_{\mathbb{Z}} (\mathbb{Z}/2\mathbb{Z}) \xrightarrow{\iota \otimes 1_{\mathbb{Z}/2\mathbb{Z}}} \mathbb{Z} \otimes_{\mathbb{Z}} (\mathbb{Z}/2\mathbb{Z})$$

is no longer injective even though $2\mathbb{Z} \xrightarrow{\iota} \mathbb{Z}$ is. (*Cf.* Exercise 3.)

In the categorical language, Example 5.3.12 shows that the tensor product is not in general a *left exact functor* in the category of R-modules, but it is indeed a *right exact functor*. It is often used to extend the base ring of a module. There are many uses for tensor products. But this book has to stop somewhere. It might as well be here. It is my sincere wish that the reader had enjoyed the glimpse of a fascinating theory.

Exercises 5.3

Throughout these exercises R denotes a ring and F denotes a field.

1. Check the proofs of Proposition 5.2.8 and of Theorem 5.2.9 to see if they work verbatim for Proposition 5.3.5 and Theorem 5.3.3 as well.

2. Let $A = (a_{ij})$ be an $n \times n$ matrix over F. Show that $\operatorname{rk} A = 1$ if and only if A is the product of an $n \times 1$ matrix and a $1 \times n$ matrix over F.

3. Let U, V and W be F-vector space and let U be a subspace of V. Show that the map $U \otimes_F W \longrightarrow V \otimes_F W$ is injective.

4. Let I be an ideal of R. Show that $(R/I) \otimes M \cong M/IM$.

5. Let M, M' be R-modules and let N, N' be submodules of M, M' respectively. Show that

$$\frac{M}{N} \otimes \frac{M'}{N'} \cong \frac{M \otimes M'}{\operatorname{Im}(N \otimes M') + \operatorname{Im}(M \otimes N')}.$$

6. Let R and S be *commutative* rings. We say S is an R-**algebra** if S is an R-module such that

$$r \cdot (s_1 s_2) = (r \cdot s_1) s_2 = s_1 (r \cdot s_2)$$

for all $r \in R$ and s_1, $s_2 \in S$.

   (a) Show that the map φ from R to S sending r to $r \cdot 1$ is a ring homomorphism.

   (b) Let $\varphi \colon R \to S$ be a ring homomorphism. Show that S is an R-algebra if we let $r \cdot s = \varphi(r)s$ for all $r \in R$ and $s \in S$.

7. Let S and T be two R-algebras. Show that $S \otimes_R T$ is also an R-algebra.

Index

255

Printed in the United States
by Baker & Taylor Publisher Services